Applied Recommender Systems with Python

Build Recommender Systems with Deep Learning, NLP and Graph-Based Techniques

Akshay Kulkarni
Adarsha Shivananda
Anoosh Kulkarni
V Adithya Krishnan

Apress®

Applied Recommender Systems with Python: Build Recommender Systems with Deep Learning, NLP and Graph-Based Techniques

Akshay Kulkarni
Bangalore, Karnataka, India

Adarsha Shivananda
Hosanagara tq, Shimoga dt, Karnataka, India

Anoosh Kulkarni
Bangalore, India

V Adithya Krishnan
Navi Mumbai, India

ISBN-13 (pbk): 978-1-4842-8953-2
https://doi.org/10.1007/978-1-4842-8954-9

ISBN-13 (electronic): 978-1-4842-8954-9

Managing Director, Apress Media LLC: Welmoed Spahr
Acquisitions Editor: Celestin Suresh John
Development Editor: Laura Berendson
Coordinating Editor: Mark Powers
Copyeditor: Kim Burton

Cover designed by eStudioCalamar

Cover image by Benyafez Studio on Unsplash (www.unsplash.com)

Distributed to the book trade worldwide by Apress Media, LLC, 1 New York Plaza, New York, NY 10004, U.S.A. Phone 1-800-SPRINGER, fax (201) 348-4505, e-mail orders-ny@springer-sbm.com, or visit www.springeronline.com. Apress Media, LLC is a California LLC and the sole member (owner) is Springer Science + Business Media Finance Inc (SSBM Finance Inc). SSBM Finance Inc is a **Delaware** corporation.

For information on translations, please e-mail booktranslations@springernature.com; for reprint, paperback, or audio rights, please e-mail bookpermissions@springernature.com.

Apress titles may be purchased in bulk for academic, corporate, or promotional use. eBook versions and licenses are also available for most titles. For more information, reference our Print and eBook Bulk Sales web page at http://www.apress.com/bulk-sales.

Any source code or other supplementary material referenced by the author in this book is available to readers on GitHub (https://github.com/Apress). For more detailed information, please visit http://www.apress.com/source-code.

Printed on acid-free paper

To our families

Table of Contents

About the Authors

Akshay R. Kulkarni is an artificial intelligence (AI) and machine learning (ML) evangelist and a thought leader. He has consulted several Fortune 500 and global enterprises to drive AI and data science–led strategic transformations. He is a Google developer, an author, and a regular speaker at major AI and data science conferences (including the O'Reilly Strata Data & AI Conference and Great International Developer Summit (GIDS)). He is a visiting faculty member at some of the top graduate institutes in India. In 2019, he was featured as one of India's "top 40 under 40" data scientists. In his spare time, Akshay enjoys reading, writing, coding, and helping aspiring data scientists. He lives in Bangalore with his family.

Adarsha Shivananda is a data science and MLOps leader. He is working on creating world-class MLOps capabilities to ensure continuous value delivery from AI. He aims to build a pool of exceptional data scientists within and outside organizations to solve problems through training programs. He always wants to stay ahead of the curve. Adarsha has worked extensively in the pharma, healthcare, CPG, retail, and marketing domains. He lives in Bangalore and loves to read and teach data science.

Anoosh Kulkarni is a data scientist and a senior AI consultant. He has worked with global clients across multiple domains to help them solve their business problems using machine learning, natural language processing (NLP), and deep learning. Anoosh is passionate about guiding and mentoring people in their data science journey. He leads data science/machine learning meet-ups and helps aspiring data scientists navigate their careers. He also conducts ML/AI workshops at universities and is actively involved in conducting webinars, talks, and sessions on AI and data science. He lives in Bangalore with his family.

V. Adithya Krishnan is a data scientist and MLOps engineer. He has worked with various global clients across multiple domains and helped them to solve their business problems using advanced ML applications. He has experience across multiple fields of AI-ML, including time-series forecasting, deep learning, NLP, ML operations, image processing, and data analytics. Presently, he is working on a state-of-the-art value observability suite for models in production, which includes continuous model and data monitoring along with the business value realized. He presented a paper, "Deep Learning Based Approach for Range Estimation," written in collaboration with the DRDO, at an IEEE conference. He lives in Chennai with his family.

About the Technical Reviewer

Krishnendu Dasgupta is co-founder of DOCONVID AI. He is a computer science and engineering graduate with a decade of experience building solutions and platforms on applied machine learning. He has worked with NTT DATA, PwC, and Thoucentric and is now working on applied AI research in medical imaging and decentralized privacy-preserving machine learning in healthcare. Krishnendu is an alumnus of the MIT Entrepreneurship and Innovation Bootcamp and devotes his free time as an applied AI and ML research volunteer for various research NGOs and universities across the world.

Preface

This book is dedicated to data scientists who are starting new recommendation engine projects from scratch but don't have prior experience in this domain. They can easily learn concepts and gain practical knowledge with this book. Recommendation engines have recently gained a lot of traction and popularity in different domains and have a proven track record for increasing sales and revenue.

This book is divided into eleven chapters. The first section, Chapters 1 and 2, covers basic approaches. The following section, which consists of Chapters 3, 4, 5, and 6, covers popular methods, including collaborative filtering-based, content-based, and hybrid recommendation systems. The next section, Chapters 7 and 8, discusses implementing systems using state-of-the-art machine learning algorithms. Chapters 9, 10, and 11 discuss trending and emerging techniques in recommendation systems.

The code for the implementations in each chapter and the required datasets are available on GitHub at `github.com/apress/applied-recommender-systems-python`.

To successfully perform all the projects in this book, you need Python 3.x or higher running on any Windows- or Unix-based operating system with a processor of 2.0 GHz or higher and a minimum of 4 GB RAM. You can download Python from Anaconda and leverage a Jupyter notebook for all coding purposes. This book assumes you know Keras basics and how to install machine learning and deep learning basic libraries.

Please upgrade or install the latest versions of all the libraries.

CHAPTER 1

Introduction to Recommendation Systems

In today's world, customers are faced with multiple choices for every decision. Let's assume that a person is looking for a book to read without any specific idea of what they want. There's a wide range of possibilities for how their search might pan out. They might waste a lot of time browsing the Internet and trawling through various sites hoping to strike gold. They might look for recommendations from other people.

But if there was a site or app that could recommend books to this customer based on what they'd read previously, that would save time that would otherwise be spent searching for books of interest on various sites. In short, our main goal is to recommend things based on the user's interests. And that's what recommendation engines do.

A *recommendation engine*, also known as a *recommender system* or a *recommendation system*, is one of the most widely used machine learning applications; for example, it is used by companies like Amazon, Netflix, Google, and Goodreads.

This chapter explains recommendation systems and presents various recommendation engine algorithms and the fundamentals of creating them in Python 3.8 or greater using a Jupyter notebook.

© Akshay Kulkarni, Adarsha Shivananda, Anoosh Kulkarni, V Adithya Krishnan 2023
A. Kulkarni et al., *Applied Recommender Systems with Python*, https://doi.org/10.1007/978-1-4842-8954-9_1

What Are Recommendation Engines?

In the past, people generally purchased products recommended to them by their friends or the people they trust. This is how people used to make purchasing decisions when there was doubt about a product. But since the advent of the Internet, we are so used to ordering online and streaming music and movies that we are constantly creating data in the back end. A recommendation engine uses that data and different algorithms to recommend the most relevant items to users. It initially captures the past behavior of a user, and then it recommends items for future purchase or use.

There are scenarios where there is no historical data as well. For example, when a new user visits a site, there is no history of that user. So how does the website recommend products to this user? One way is by recommending bestselling products (i.e., the products that are trending). Another possible solution is to recommend the products that bring maximum profit to the business and any new products recently added to the site.

If you can recommend a few items to a customer based on their interests, it positively impacts the user experience and leads to frequent visits. Hence, intelligent recommendation engines are built by studying the past behavior of their users to enhance revenue.

Recommendation System Types

Data on a user's likes and dislikes of items are essential to building a recommender engine that can suggest relevant items to the user. There are two feedback mechanisms through which users provide this required data.

Explicit feedback is the data that the user explicitly provides as feedback on an item. It is usually difficult to obtain this type of feedback from users, and companies try many innovative ways. A simple like or dislike button, star ratings, and even comments and reviews as text input can get user feedback on an item.

Implicit feedback is the data that the user implicitly or unknowingly provides through their actions. This can be in the form of pages visited, items viewed, the number of clicks, and all sorts of other activities performed on the site/platform, which can indicate their interest in certain items. This type of data is generally captured automatically through cookies and browsing history and doesn't require any direct action from the users.

Types of Recommendation Engines

There are many different types of recommendation engines, and each of them is explored in this chapter.

- Market basket analysis (association rule mining)
- Content-based filtering
- Collaborative-based filtering
- Hybrid systems
- ML clustering
- ML classification
- Deep learning and NLP

Market Basket Analysis (Association Rule Mining)

Retailers predominantly use *market basket analysis* to reveal relationships between items. It works by looking for combinations of items that are often put together, allowing retailers to identify relationships between items that people buy.

There are several terms used in association analysis that are important to understand. Association rules are widely used to analyze retail basket or transaction data. They are intended to identify strong rules discovered in transaction data using interest measures based on the concept of strong rules.

Association rules are normally written like this: {bread} -> {butter}. This means a strong relationship exists between customers who bought bread and butter in the same transaction.

In the preceding example, {bread} is the antecedent and {butter} is the consequent. Both antecedents and consequences can have multiple items. In other words, {bread, milk} -> {butter, chips} is a valid rule.

Support is the relative frequency of the rule display. In many cases, you may want to seek high support to make sure it's a worthwhile relationship. However, there may be cases where low support is useful if you are trying to find "hidden" relationships.

Confidence is a measure of the reliability of a rule. A 0.5 reliability in the preceding example means that bread and milk were purchased 50% of the time. The purchase also included butter and chips. For a product recommendation, 50% confidence may be perfectly acceptable, but this level may not be high enough in a medical situation.

Lift is the ratio of observed support to expected support if the two rules were independent. As a rule of thumb, a lift value close to 1 means that the rules were completely independent. Lift - values > 1 are more "interesting" and could indicate a useful rule pattern. Figure 1-1 illustrates how support, confidence, and lift are calculated.

Rule	Support	Confidence	Lift
$A \Rightarrow D$	2/5	2/3	10/9
$C \Rightarrow A$	2/5	2/4	5/6
$A \Rightarrow C$	2/5	2/3	5/6
$B \& C \Rightarrow D$	1/5	1/3	5/9

Figure 1-1. *Market basket analysis*

Content-Based Filtering

The content-based filtering method is a recommendation algorithm that suggests items similar to the ones the users have previously selected or shown interest in. It can recommend based on the actual content present in the item. For example, as shown in Figure 1-2, a new article is recommended based on the text present in the articles.

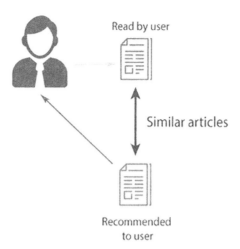

Figure 1-2. *Content-based system*

Let's look at the popular example of Netflix and its recommendations to explore the workings in detail. Netflix saves all user viewing information in a vector-based format, known as the *profile vector*, which contains information on past viewings, liked and disliked shows, most frequently watched genres, star ratings, and so forth. Then there is another vector that stores all the information regarding the titles (movies and shows) available on the platform, known as the item vector. This vector stores information like the title, actors, genre, language, length, crew info, synopsis, and so forth.

The content-based filtering algorithm uses the concept of cosine similarity. In it, you find the cosine of the angle between two vectors—the profile and item vectors in this case. Suppose *A* is the profile vector and *B* is the item vector, then the (cosine) similarity between them is calculated as follows.

$$sim(A, B) = \cos(\theta) = \frac{A \cdot B}{\|A\|\|B\|}$$

The outcome (i.e., the cosine value) always ranges between –1 and 1, and this value is calculated for multiple item vectors (movies), keeping the profile vector (user) constant. The items/movies are then ranked in descending order of similarity, and either of the two following approaches is used for recommendations.

- In a **top N approach**, the top N movies are recommended, where N is a threshold on the number of titles recommended.

- In a **rating scale approach**, a threshold on the similarity value is set, and all the titles in that threshold are recommended.

The following are other methods popularly used in calculating the similarity.

- **Euclidean distance** is the distance between two points measured by the length of the straight line connecting them. Hence if you can plot the profile and items in an n-dimensional Euclidean space, the similarity value is equal to the distance between them. The closer the item is, the more similar it is. So, the closest items to the profile are recommended. The following is the mathematical formula for calculating Euclidean distance.

$$\text{Euclidean Distance} = \sqrt{(x_1 - y_1)^2 + \ldots + (x_N - y_N)^2}$$

- *Pearson's correlation* refers to how correlated or similar two things are. The higher the correlation, the higher the similarity. Pearson's correlation is calculated using the formula shown in Figure 1-3.

$$sim(u, v) = \frac{\sum (r_{ui} - \bar{r}_u)(r_{vi} - \bar{r}_v)}{\sqrt{\sum (r_{ui} - \bar{r}_u)^2} \sqrt{\sum (r_{vi} - \bar{r}_v)^2}}$$

Figure 1-3. *Formula*

The major downside to this recommendation engine is that all suggestions fall into the same category, and it becomes somewhat monotonous. As the suggestions are based on what the user has seen or liked, we'll never get new recommendations that the user has not explored in the past. For example, if the user has only seen mystery movies, the engine will only recommend more mystery movies.

To improve on this, you need a recommendation engine that not only gives suggestions based on the content but also on the behavior of users and on what other like-minded users are watching.

Collaborative-Based Filtering

In collaborative-based filtering recommendation engines, a user-to-user similarity is also considered, along with item similarities, to address some of the drawbacks of content-based filtering. Simply put, a collaborative filtering system recommends an item to user A based on the interests of a similar user B. Figure 1-4 shows a simple working mechanism of collaborative-based filtering

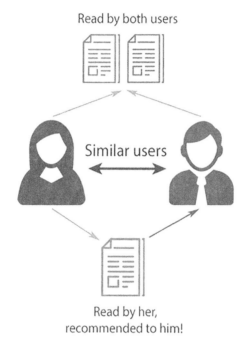

Figure 1-4. *Collaborative-based filtering*

The similarity between users can be calculated again by all the techniques mentioned earlier. A user-item matrix is created individually for each customer, which stores the user's preference for an item. Taking the same example of Netflix's recommendation engine, the user aspects like previously watched and liked titles, ratings provided (if any) by the user, frequently watched genres, and so on are stored and used to find similar users. Once these similar users are found, the engine recommends titles that the user has not yet watched but users with similar interests have watched and liked.

This type of filtering is quite popular because it is only based on a user's past behavior, and no additional input is required. It's used by many major companies, including Amazon, Netflix, and American Express.

There are two types of collaborative filtering algorithms.

- In **user-user collaborative filtering**, you find user-user similarities and offer suggestions based on what similar users chose in the past. Even though this algorithm is quite effective, since it requires high computations for getting all user-pair information and calculating the similarities, it takes a lot of time and resources. Hence for big customer bases, this algorithm is too expensive to use unless a proper parallelizable system is set up.

- In **item-item collaborative filtering**, you try to find item similarities instead of similar users. An item look-alike matrix is generated for all the items that the user has previously chosen, and from this matrix, similar items are recommended. This algorithm is far less computationally expensive because the item-item look-alike matrix remains fixed over time with a fixed number of items. Hence recommendations are fetched much quicker for a new customer.

One of the drawbacks of this method happens when no ratings are provided for a particular item; then, it can't be recommended. And reliable recommendations can be tough to get if a user has only rated a few items.

Hybrid Systems

So far, you have seen how content-based and collaborative-based recommendation engines work and their respective pros and cons. But the *hybrid recommendation system* combines content and collaborative-based filtering methods.

Hybrid recommendation systems can overcome the drawbacks of both content-based and collaborative-based to form one powerful recommendation system, both the individual methods fail to perform well when there is a lack of data to learn the relation between users and items, which is overcome in this hybrid approach.

Figure 1-5 shows a simple working mechanism of the hybrid recommendation system.

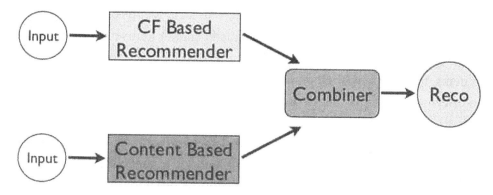

Figure 1-5. *Hybrid recommendation system*

Hybrid recommendation engines can be implemented in multiple ways.

- Generating recommendations separately by using content-based and collaborative-based and then combining them at the end

- Adding the capabilities of the collaborative-based method to a content-based recommender engine

- Adding the capabilities of the content-based method to a collaborative-based recommender engine

Several studies compare the performance of conventional methods to that of a hybrid system, showing that hybrid recommender engines generally perform better and provide more reliable recommendations.

ML Clustering

In today's world, AI has become an integral part of all automation and technology-based solutions and the area of recommendation systems is no different. Machine learning-based methods are the upcoming high prospective methods that are quickly becoming a norm as more and more companies start adapting AI.

Machine learning methods are of two types: unsupervised and supervised. This section discusses the unsupervised learning method, which is the ML clustering–based method. The unsupervised learning technique uses ML algorithms to find hidden patterns in data to cluster them without human intervention (unlabeled data). Clustering is the grouping of similar objects into clusters. On average, an object belonging to one cluster is more similar to an object within that cluster than to an object belonging to another cluster.

In recommendation engines, clustering is used to form groups of users similar to each other, as shown in Figure 1-6. It can also cluster similar items or products as well. Traditionally similarity measures like cosine similarity have been used to get similar users or items, but they have their demerits. If a user has not rated many items and the resultant user-item matrix is sparse, or when you need to compare multiple user-user pairings to find similar users, it gets computationally expensive. A clustering-based approach is generally taken to get similar users to overcome these issues. If a user is found to be similar to a cluster of users, that user is added to that cluster. Within the cluster, all users share interests and tastes, and recommendations are provided to users based on them.

Figure 1-6. *Groups based on behavior*

The following are some of the popularly used clustering algorithms.

- k-means clustering
- fuzzy mapping
- self-organizing maps (SOM)
- a hybrid of two or more techniques

ML Classification

Again, clustering comes with its disadvantages. That's where a classification-based recommendation system comes into play.

In classification based, the algorithm uses features of both items and users to predict whether a user will like a product or not. An application of the classification-based method is the buyer propensity model.

Propensity modeling predicts the chances of customers buying a particular item or any equivalent task. Also, for example, propensity modeling can help predict the likelihood that a sales lead will convert to a customer or not based on various features. The propensity score or probability is used to take action.

The following are some of the limitations of classification-based algorithms.

- Collecting a combination of data about different users and items is sometimes difficult.

- Classification is challenging.

- The problem is training the models in real time.

Deep Learning

Deep learning is a branch of machine learning which is more powerful than ML-based algorithms and tends to produce better results. Of course, there are limitations, like the need for huge data or explainability, which we must overcome.

Various companies use deep neural networks (DNNs) to enhance the customer experience, especially if it's unstructured data like images and text.

The following are three types of deep learning–based recommender systems.

- Restricted Boltzmann

- Autoencoder based

- Neural attention–based

Later chapters explore how machine learning and deep learning can be leveraged to build powerful recommender systems.

Now that you have a good understanding of the concepts, let's start with a simple rule-based recommender system in this chapter before proceeding to the implementation in upcoming chapters.

Rule-Based Recommendation Systems

You build these recommendation systems with simple rules, such as popularity-based or buy again.

Popularity

A popularity-based rule is the simplest form: a product is recommended based on its popularity (most sold, most clicked, etc.). Let's implement a quick one. For example, a song listened to by many people means it's popular. It is recommended to others without any other intelligence being part of it.

Let's take a retail dataset and implement the same logic.

Fire up a Jupyter notebook and import the necessary packages.

```
#import necessary libraries
import pandas as pd
import numpy as np

#import viz libraries
import seaborn as sns
import matplotlib.pyplot as plt
%matplotlib inline
```

Let's import the data.

Note   Refer to the data in this book's data section. Download the dataset from the GitHub link of this book.

```
#import data
df = pd.read_csv('data.csv',encoding= 'unicode_escape')
df.head()
```

Figure 1-7 shows the output of the top 5 rows from the dataset.

	InvoiceNo	StockCode	Description	Quantity	InvoiceDate	UnitPrice	CustomerID	Country
0	536365	85123A	WHITE HANGING HEART T-LIGHT HOLDER	6	12/1/2010 8:26	2.55	17850.0	United Kingdom
1	536365	71053	WHITE METAL LANTERN	6	12/1/2010 8:26	3.39	17850.0	United Kingdom
2	536365	84406B	CREAM CUPID HEARTS COAT HANGER	8	12/1/2010 8:26	2.75	17850.0	United Kingdom
3	536365	84029G	KNITTED UNION FLAG HOT WATER BOTTLE	6	12/1/2010 8:26	3.39	17850.0	United Kingdom
4	536365	84029E	RED WOOLLY HOTTIE WHITE HEART.	6	12/1/2010 8:26	3.39	17850.0	United Kingdom

Figure 1-7. *The output*

```
# null value counts
df.isnull().sum().sort_values(ascending=False)
```

```
CustomerID      135080
Description       1454
Country              0
UnitPrice            0
InvoiceDate          0
Quantity             0
StockCode            0
InvoiceNo            0
dtype: int64
```

```
# drop where Description is not available
df_new = df.dropna(subset=['Description'])
df_new.describe()
```

Figure 1-8 shows that the quantity has some negative values that are a part of the incorrect data, so lets drop them using the below code.

```
df_new = df_new[df_new.Quantity > 0]
df_new.describre()
```

	Quantity	UnitPrice	CustomerID
count	540455.000000	540455.000000	406829.000000
mean	9.603129	4.623519	15287.690570
std	218.007598	96.889628	1713.600303
min	-80995.000000	-11062.060000	12346.000000
25%	1.000000	1.250000	13953.000000
50%	3.000000	2.080000	15152.000000
75%	10.000000	4.130000	16791.000000
max	80995.000000	38970.000000	18287.000000

Figure 1-8. *The output contains negative values*

	Quantity	UnitPrice	CustomerID
count	530693.000000	530693.000000	397924.000000
mean	10.605819	3.861599	15294.315171
std	156.637853	41.833162	1713.169877
min	1.000000	-11062.060000	12346.000000
25%	1.000000	1.250000	13969.000000
50%	3.000000	2.080000	15159.000000
75%	10.000000	4.130000	16795.000000
max	80995.000000	13541.330000	18287.000000

Figure 1-9. *shows the output after removing the negative values*

Now that we cleaned up the data, let's do some basic types of recommendation systems. These are not intelligent yet effective in some cases. Popularity-based recommendation systems could be a trending song. It could be a fast-selling item required for everyone, a recently released movie that gets traction, or a news article many users have read.

Sometimes it's important to keep it simple because it gets you the most revenue. Let's build a popularity-based system in the data we are using.

Global Popular Items

Let's calculate popular items worldwide and then dice them into different regions.

```
# popular items globally
global_popularity=df_new.pivot_table(index=['StockCode','Description'],
values='Quantity', aggfunc='sum').sort_values(by='Quantity',
ascending=False)
print('Top 10 popular items globally....')
global_popularity.head(10)
```

Figure 1-10 shows that PAPER CRAFT is the most bought item across all regions. It's a very popular item.

```
Top 10 popular items globally....
```

StockCode	Description	Quantity
23843	PAPER CRAFT , LITTLE BIRDIE	80995
23166	MEDIUM CERAMIC TOP STORAGE JAR	78033
84077	WORLD WAR 2 GLIDERS ASSTD DESIGNS	55047
85099B	JUMBO BAG RED RETROSPOT	48478
85123A	WHITE HANGING HEART T-LIGHT HOLDER	37603
22197	POPCORN HOLDER	36761
84879	ASSORTED COLOUR BIRD ORNAMENT	36461
21212	PACK OF 72 RETROSPOT CAKE CASES	36419
23084	RABBIT NIGHT LIGHT	30788
22492	MINI PAINT SET VINTAGE	26633

Figure 1-10. *The output*

Let's visualize it.

```
# vizualize top 10 most popular items

global_popularity.reset_index(inplace=True)
sns.barplot(y='Description', x='Quantity', data=global_popularity.head(10))
plt.title('Top 10 Most Popular Items Globally', fontsize=14)
plt.ylabel('Item')
```

Figure 1-11 shows the output of top 10 popular items.

```
Text(0, 0.5, 'Item')
```

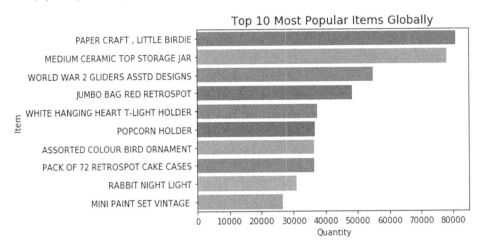

Figure 1-11. *The output*

Popular Items by Country

Let's calculate popular items by country.

```
# popular items by country
countrywise=df_new.pivot_table(index=['Country','StockCode','Description'],
values='Quantity', aggfunc='sum').reset_index()
```

```
# vizualize top 10 most popular items in UK
sns.barplot(y='Description', x='Quantity', data=countrywise[countrywise
['Country']=='United Kingdom'].sort_values(by='Quantity', ascending=False).
head(10))
plt.title('Top 10 Most Popular Items in UK', fontsize=14)
plt.ylabel('Item')
```

Figure 1-12 shows that PAPER CRAFT, LITTLE BIRDIE is the most purchased item. It's very popular only in the United Kingdom.

```
# vizualize top 10 most popular items in Netherlands

sns.barplot(y='Description', x='Quantity', data=countrywise[countrywise
['Country']=='Netherlands'].sort_values(by='Quantity', ascending=False).
head(10))
plt.title('Top 10 Most Popular Items in Netherlands', fontsize=14)
plt.ylabel('Item')
```

```
Text(0, 0.5, 'Item')
```

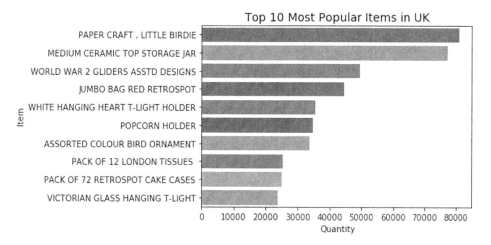

Figure 1-12. *The output*

Figure 1-13 shows that RABBIT NIGHT LIGHT is the most purchased item. It's very popular in the Netherlands.

```
Text(0, 0.5, 'Item')
```

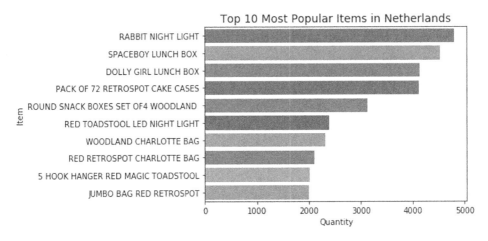

Figure 1-13. *The output*

Buy Again

Now let's discuss *buy again*. It's another simple recommendation simple calculated at the customer/user level. You might have seen "Watch again" on streaming platforms. It's the same concept. You know a certain set of actions are done repeatedly by a customer, and we recommend the same action next time.

This is very useful in online grocery platforms because customers come back and buy the same item again and again.

Let's implement it.

```
# Lets create a function to get buy again output
from collections import Counter
def buy_again(customerid):

    # Fetching the items bought by the customer for provided customer id
    items_bought = df_new[df_new['CustomerID']==customerid].Description

    # Count and sort the repeated purchases
    bought_again = Counter(items_bought)

    # Convert counter to list for printing recommendations
    buy_again_list = list(bought_again)
```

```
# Printing the recommendations
print('Items you would like to buy again :')
return(buy_again_list)
```

Let's use the function on customer 17850.

```
buy_again(17850)
```

Figure 1-14 recommends the holder and the lantern to customer 17850, given that he often buys these items.

```
Items you would like to buy again :

['WHITE HANGING HEART T-LIGHT HOLDER',
 'WHITE METAL LANTERN',
 'CREAM CUPID HEARTS COAT HANGER',
 'KNITTED UNION FLAG HOT WATER BOTTLE',
 'RED WOOLLY HOTTIE WHITE HEART.',
 'SET 7 BABUSHKA NESTING BOXES',
 'GLASS STAR FROSTED T-LIGHT HOLDER',
 'HAND WARMER UNION JACK',
 'HAND WARMER RED POLKA DOT',
 'EDWARDIAN PARASOL RED',
 'RETRO COFFEE MUGS ASSORTED',
 'SAVE THE PLANET MUG',
 'VINTAGE BILLBOARD DRINK ME MUG',
 'VINTAGE BILLBOARD LOVE/HATE MUG',
 'WOOD 2 DRAWER CABINET WHITE FINISH',
 'WOOD S/3 CABINET ANT WHITE FINISH',
 'WOODEN PICTURE FRAME WHITE FINISH',
 'WOODEN FRAME ANTIQUE WHITE ',
 'EDWARDIAN PARASOL BLACK',
 'IVORY EMBROIDERED QUILT ',
 'JUMBO SHOPPER VINTAGE RED PAISLEY']
```

Figure 1-14. *The output*

Summary

In this chapter, you learned about recommender systems—how they work, their applications, and the various implementation types. You also learned about implicit and explicit types and the differences between them. The chapter also explored market basket analysis (association rule mining), content-based and collaborative-based filtering, hybrid systems, ML clustering-based and classification-based methods, and deep learning and NLP-based recommender systems. Finally, you implemented simple recommender systems. Many other complex algorithms are explored in upcoming chapters.

CHAPTER 2

Market Basket Analysis (Association Rule Mining)

Market basket analysis (MBA) is a technique used in data mining by retail companies to increase sales by better understanding customer buying patterns. It involves analyzing large datasets, such as customer purchase history, to uncover item groupings and products that are likely to be frequently purchased together.

Figure 2-1 explains the MBA at a high level.

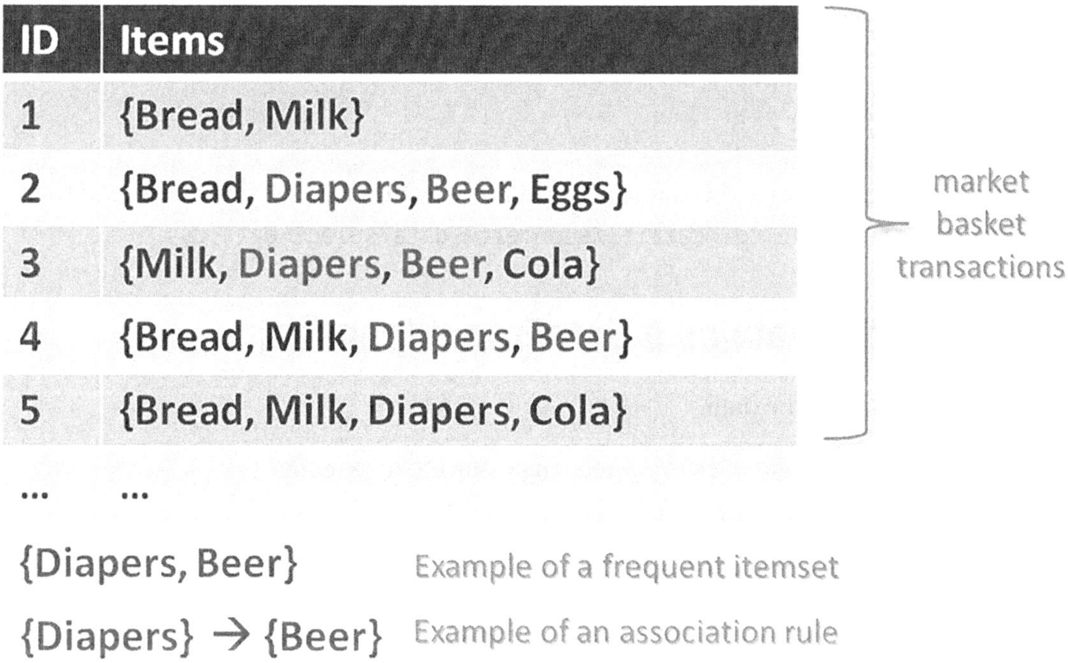

ID	Items	
1	{Bread, Milk}	
2	{Bread, Diapers, Beer, Eggs}	market
3	{Milk, Diapers, Beer, Cola}	basket
4	{Bread, Milk, Diapers, Beer}	transactions
5	{Bread, Milk, Diapers, Cola}	
...	...	

{Diapers, Beer}   Example of a frequent itemset

{Diapers} → {Beer}   Example of an association rule

Figure 2-1. *MBA explained*

This chapter explores the implementation of market basket analysis with the help of an open source e-commerce dataset. You start with the dataset in *exploratory data*

© Akshay Kulkarni, Adarsha Shivananda, Anoosh Kulkarni, V Adithya Krishnan 2023
A. Kulkarni et al., *Applied Recommender Systems with Python*, https://doi.org/10.1007/978-1-4842-8954-9_2

analysis (EDA) and focus on critical insights. You then learn about the implementation of various techniques in MBA, plot a graphical representation of the associations, and draw insights.

Implementation

Let's imports the required libraries.

```
import pandas as pd
import numpy as np
import seaborn as sns
import matplotlib.pyplot as plt
import matplotlib.style
%matplotlib inline
from mlxtend.frequent_patterns import apriori,association_rules
from collections import Counter
from IPython.display import Image
```

Data Collection

Let's look at an open source dataset from a Kaggle e-commerce website. Download the dataset from www.kaggle.com/carrie1/ecommerce-data?select=data.csv.

Importing the Data as a DataFrame (pandas)

The following imports the data.

```
data = pd.read_csv('data.csv', encoding= 'unicode_escape')
data.shape
```

The following is the output.

```
(541909, 8)
```

Let's print the top five rows of the DataFrame.

```
data.head()
```

Figure 2-2 shows the output of the first five rows.

	InvoiceNo	StockCode	Description	Quantity	InvoiceDate	UnitPrice	CustomerID	Country
0	536365	85123A	WHITE HANGING HEART T-LIGHT HOLDER	6	12/1/2010 8:26	2.55	17850.0	United Kingdom
1	536365	71053	WHITE METAL LANTERN	6	12/1/2010 8:26	3.39	17850.0	United Kingdom
2	536365	84406B	CREAM CUPID HEARTS COAT HANGER	8	12/1/2010 8:26	2.75	17850.0	United Kingdom
3	536365	84029G	KNITTED UNION FLAG HOT WATER BOTTLE	6	12/1/2010 8:26	3.39	17850.0	United Kingdom
4	536365	84029E	RED WOOLLY HOTTIE WHITE HEART.	6	12/1/2010 8:26	3.39	17850.0	United Kingdom

Figure 2-2. *The output*

Check for nulls in the data.

```
data.isnull().sum().sort_values(ascending=False)
```

The following is the output.

```
CustomerID      135080
Description       1454
Country              0
UnitPrice            0
InvoiceDate          0
Quantity             0
StockCode            0
InvoiceNo            0
dtype: int64
```

Cleaning the Data

The following drops nulls and describes the data.

```
data1 = data.dropna()
data1.describe()
```

Figure 2-3 shows the output after dropping nulls.

23

	Quantity	UnitPrice	CustomerID
count	406829.000000	406829.000000	406829.000000
mean	12.061303	3.460471	15287.690570
std	248.693370	69.315162	1713.600303
min	-80995.000000	0.000000	12346.000000
25%	2.000000	1.250000	13953.000000
50%	5.000000	1.950000	15152.000000
75%	12.000000	3.750000	16791.000000
max	80995.000000	38970.000000	18287.000000

Figure 2-3. *The output*

The Quantity column has some negative values, which are part of the incorrect data, so let's drop these entries.

The following selects only data in which the quantity is greater than 0.

```
data1 = data1[data1.Quantity > 0]
data1.describe()
```

Figure 2-4 shows the output after filtering the data in the Quantity column.

	Quantity	UnitPrice	CustomerID
count	397924.000000	397924.000000	397924.000000
mean	13.021823	3.116174	15294.315171
std	180.420210	22.096788	1713.169877
min	1.000000	0.000000	12346.000000
25%	2.000000	1.250000	13969.000000
50%	6.000000	1.950000	15159.000000
75%	12.000000	3.750000	16795.000000
max	80995.000000	8142.750000	18287.000000

Figure 2-4. *The output*

Insights from the Dataset

Customer Insights

This segment answers the following questions.

- Who are my loyal customers?

- Which customers have ordered most frequently?

- Which customers contribute the most to my revenue?

Loyal Customers

Let's create a new Amount feature/column, which is the product of the quantity and its unit price.

```
data1['Amount'] = data1['Quantity'] * data1['UnitPrice']
```

Now let's use the group by function to highlight the customers with the greatest number of orders.

```
orders = data1.groupby(by=['CustomerID','Country'], as_index=False)
['InvoiceNo'].count()

print('The TOP 5 loyal customers with the most number of orders...')
orders.sort_values(by='InvoiceNo', ascending=False).head()
```

Figure 2-5 shows the top five loyal customers.

```
The TOP 5 loyal customers with most number of orders...
```

	CustomerID	Country	InvoiceNo
4019	17841.0	United Kingdom	7847
1888	14911.0	EIRE	5677
1298	14096.0	United Kingdom	5111
334	12748.0	United Kingdom	4596
1670	14606.0	United Kingdom	2700

Figure 2-5. *The output*

Number of Orders per Customer

Let's plot the orders by different customers.
    Create a subplot of size 15×6.

```
plt.subplots(figsize=(15,6))
```

Use bmh for better visualization.

```
plt.style.use('bmh')
```

The x axis indicates the customer ID, and the y axis indicates the number of orders.

```
plt.plot(orders.CustomerID, orders.InvoiceNo)
```

Let's label the x axis and the y axis.

```
plt.xlabel('Customers ID')
plt.ylabel('Number of Orders')
```

Give a suitable title to the plot.

```
plt.title('Number of Orders by different Customers')
plt.show()
```

Figure 2-6 shows the number of orders by different customers.

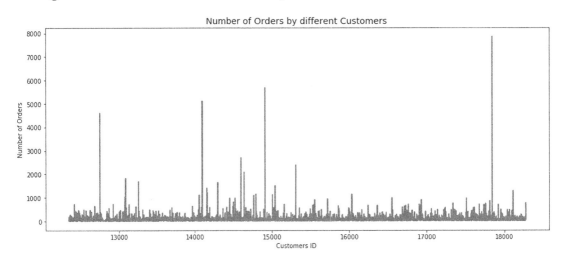

Figure 2-6. *The output*

Let's use the group by function again to get the customers with the highest amount spent (invoices).

```
money_spent = data1.groupby(by=['CustomerID','Country'], as_index=False)
['Amount'].sum()
```

```
print('The TOP 5 profitable customers with the highest money spent...')
money_spent.sort_values(by='Amount', ascending=False).head()
```

Figure 2-7 shows the top five profitable customers.

```
Out[24]:
            CustomerID        Country    Amount
     1711    14646.0      Netherlands  279489.02
     4241    18102.0   United Kingdom  256438.49
     3766    17450.0   United Kingdom  187482.17
     1903    14911.0             EIRE  132572.62
       57    12415.0        Australia  123725.45
```

Figure 2-7. *The output*

Money Spent per Customer

Create a subplot of size 15×6.

```
plt.subplots(figsize=(15,6))
```

The x axis indicates the customer ID, and y axis indicates the amount spent.

```
plt.plot(money_spent.CustomerID, money_spent.Amount)
```

Let's use bmh for better visualization.

```
plt.style.use('bmh')
```

The following labels the x axis and the y axis.

```
plt.xlabel('Customers ID')
plt.ylabel('Money spent')
```

Let's give a suitable title to the plot.

```
plt.title('Money Spent by different Customers')
plt.show()
```

Figure 2-8 shows money spent by different customers.

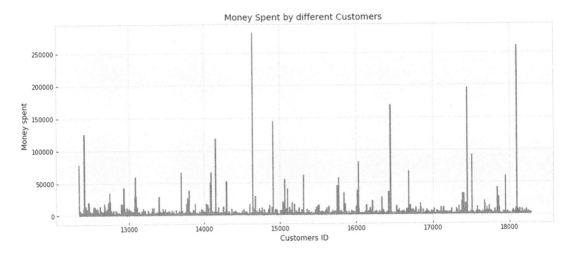

Figure 2-8. *The output*

Patterns Based on DateTime

This segment answers questions like the following.

- In which month is the highest number of orders placed?

- On which day of the week is the highest number of orders placed?

- At what time of the day is the store the busiest?

Preprocessing the Data

The following imports the DateTime library.

```
import datetime
```

The following converts InvoiceDate from an object to a DateTime format.

```
data1['InvoiceDate'] = pd.to_datetime(data1.InvoiceDate,
format='%m/%d/%Y %H:%M')
```

Let's create a new feature using the month and year.

```
data1.insert(loc=2, column='year_month', value=data1['InvoiceDate'].
map(lambda x: 100*x.year + x.month))
```

Create a new feature for the month.

```
data1.insert(loc=3, column='month', value=data1.InvoiceDate.dt.month)
```

Create a new feature for the day; for example, Monday=1.....until Sunday=7.

```
data1.insert(loc=4, column='day', value=(data1.InvoiceDate.dt.dayofweek)+1)
```

Create a new feature for the hour.

```
data1.insert(loc=5, column='hour', value=data1.InvoiceDate.dt.hour)
```

How Many Orders Are Placed per Month?

Use bmh style for better visualization.

```
plt.style.use('bmh')
```

Let's use group by to extract the number of invoices per year and month.

```
ax = data1.groupby('InvoiceNo')['year_month'].unique().value_counts().sort_
index().plot(kind='bar',figsize=(15,6))
```

The following labels the x axis and the y axis.

```
ax.set_xlabel('Month',fontsize=15)
ax.set_ylabel('Number of Orders',fontsize=15)
```

Let's give a suitable title to the plot.

```
ax.set_title(' # orders for various months (Dec 2010 - Dec
2011)',fontsize=15)
```

Provide X tick labels.

```
ax.set_xticklabels(('Dec_10','Jan_11','Feb_11','Mar_11','Apr_11','May_1
1','Jun_11','July_11','Aug_11','Sep_11','Oct_11','Nov_11','Dec_11'),
rotation='horizontal', fontsize=13)
plt.show()
```

Figure 2-9 shows the number of orders in different months.

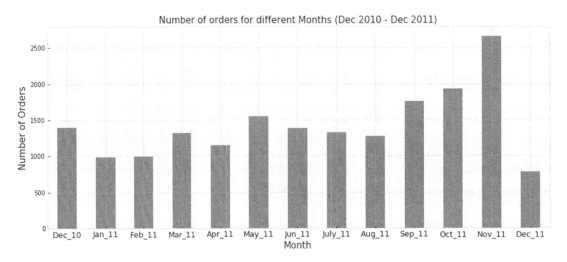

Figure 2-9. *The output*

How Many Orders Are Placed per Day?

Day = 6 is Saturday; there are no orders placed on Saturdays.

```
data1[data1['day']==6].shape[0]
```

Let's use groupby to count the number of invoices by day.

```
ax = data1.groupby('InvoiceNo')['day'].unique().value_counts().sort_
index().plot(kind='bar',figsize=(15,6))
```

The following labels the x axis and the y axis.

```
ax.set_xlabel('Day',fontsize=15)
ax.set_ylabel('Number of Orders',fontsize=15)
```

Let's give a suitable title to the plot.

```
ax.set_title('Number of orders for different Days',fontsize=15)
```

Provide X tick labels.

Since no orders were placed on Saturdays, it is excluded from xticklabels.

```
ax.set_xticklabels(('Mon','Tue','Wed','Thur','Fri','Sun'),
rotation='horizontal', fontsize=15)
plt.show()
```

Figure 2-10 shows the number of orders for different days.

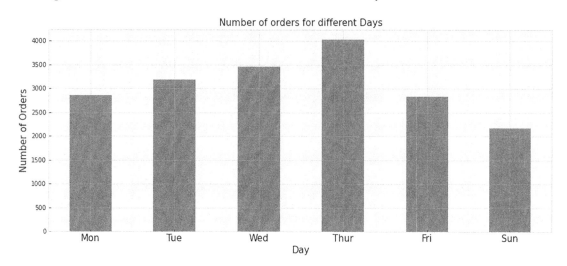

Figure 2-10. *The output*

How Many Orders Are Placed per Hour?

Let's use groupby to count the number of invoices by the hour.

```
ax = data1.groupby('InvoiceNo')['hour'].unique().value_counts().iloc[:-1].
sort_index().plot(kind='bar',figsize=(15,6))
```

The following labels the x axis and the y axis.

```
ax.set_xlabel('Hour',fontsize=15)
ax.set_ylabel('Number of Orders',fontsize=15)
```

Give a suitable title to the plot.

```
ax.set_title('Number of orders for different Hours',fontsize=15)
```

Provide X tick labels (all orders are placed between hours 6 and 20).

```
ax.set_xticklabels(range(6,21), rotation='horizontal', fontsize=15)
plt.show()
```

Figure 2-11 shows the number of orders for different hours.

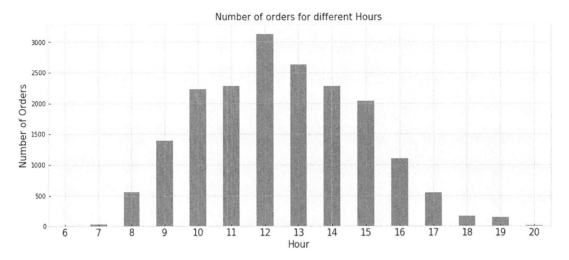

Figure 2-11. *The output*

Free Items and Sales

This segment displays how "free" items impact the number of orders. It answers how discounts and other offers impact sales.

```
data1.UnitPrice.describe()
```

The following is the output.

```
count     397924.000000
mean           3.116174
std           22.096788
min            0.000000
25%            1.250000
50%            1.950000
75%            3.750000
max         8142.750000
Name: UnitPrice, dtype: float64
```

Since the minimum unit price = 0, there are either incorrect entries or free items. Let's check the distribution of unit prices.

```
plt.subplots(figsize=(12,6))
```

Use the darkgrid style for better visualization.

```
sns.set_style('darkgrid')
```

Apply boxplot visualization to the unit price.

```
sns.boxplot(data1.UnitPrice)
plt.show()
```

Figure 2-12 shows the boxplot for unit price.

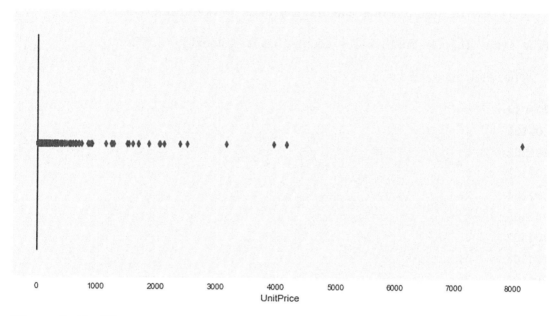

Figure 2-12.   *The output*

Items with UnitPrice = 0 are not outliers. These are the "free" items.
Create a new DataFrame for free items.

```
free_items_df = data1[data1['UnitPrice'] == 0]
free_items_df.head()
```

Figure 2-13 shows the filtered data output (unit price = 0).

	InvoiceNo	StockCode	year_month	month	day	hour	Description	Quantity	InvoiceDate	UnitPrice	CustomerID	Country	Amount
9302	537197	22841	201012	12	7	14	ROUND CAKE TIN VINTAGE GREEN	1	2010-12-05 14:02:00	0.0	12647.0	Germany	0.0
33576	539263	22580	201012	12	4	14	ADVENT CALENDAR GINGHAM SACK	4	2010-12-16 14:36:00	0.0	16560.0	United Kingdom	0.0
40089	539722	22423	201012	12	2	13	REGENCY CAKESTAND 3 TIER	10	2010-12-21 13:45:00	0.0	14911.0	EIRE	0.0
47068	540372	22090	201101	1	4	16	PAPER BUNTING RETROSPOT	24	2011-01-06 16:41:00	0.0	13081.0	United Kingdom	0.0
47070	540372	22553	201101	1	4	16	PLASTERS IN TIN SKULLS	24	2011-01-06 16:41:00	0.0	13081.0	United Kingdom	0.0

Figure 2-13. *The output*

Let's count the number of free items given away by month and year.

```
free_items_df.year_month.value_counts().sort_index()
```

The following is the output.

```
201012      3
201101      3
201102      1
201103      2
201104      2
201105      2
201107      2
201108      6
201109      2
201110      3
201111     14
Name: year_month, dtype: int64
```

There is at least one free item every month except June 2011.

Let's count the number of free items per year and month.

```
ax = free_items_df.year_month.value_counts().sort_index().plot(kind='bar',
figsize=(12,6))
```

Let's label the x axis and the y axis.

```
ax.set_xlabel('Month',fontsize=15)
ax.set_ylabel('Frequency',fontsize=15)
```

Give a suitable title to the plot.

```
ax.set_title('Frequency for different Months (Dec 2010 - Dec
2011)',fontsize=15)
```

Provide X tick labels.

Since there were no free items in June 2011, it is excluded.

```
ax.set_xticklabels(('Dec_10','Jan_11','Feb_11','Mar_11','Apr_11','May_11
','July_11','Aug_11','Sep_11','Oct_11','Nov_11'), rotation='horizontal',
fontsize=13)
plt.show()
```

Figure 2-14 shows the frequency for different months.

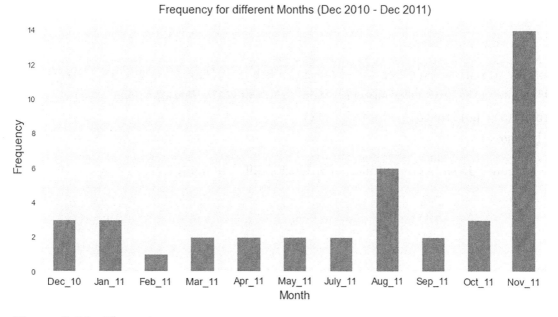

Figure 2-14. *The output*

The greatest number of free items were given out in November 2011. The greatest number of orders were also placed in November 2011.

Use bmh.

```
plt.style.use('bmh')
```

Use groupby to count the unique number of invoices by year and month.

```
ax = data1.groupby('InvoiceNo')['year_month'].unique().value_counts().sort_
index().plot(kind='bar',figsize=(15,6))
```

The following labels the x axis.

```
ax.set_xlabel('Month',fontsize=15
```

The following labels the y axis.

```
ax.set_ylabel('Number of Orders',fontsize=15)
```

Give a suitable title to the plot.

```
ax.set_title('# Number of orders for different Months (Dec 2010 - Dec
2011)',fontsize=15)
```

Provide X tick labels.

```
ax.set_xticklabels(('Dec_10','Jan_11','Feb_11','Mar_11','Apr_11','May_1
1','Jun_11','July_11','Aug_11','Sep_11','Oct_11','Nov_11','Dec_11'),
rotation='horizontal', fontsize=13)
plt.show()
```

Figure 2-15 shows the number of orders for different months.

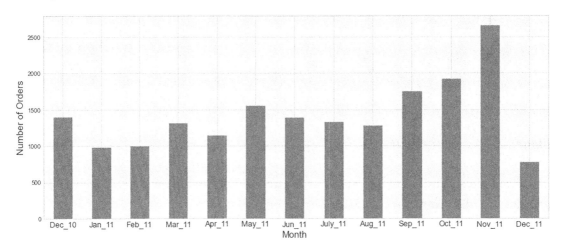

Figure 2-15. *The output*

Compared to the May month, the sales for the month of August have declined, indicating a slight effect from the "number of free items".

Use bmh.

```
plt.style.use('bmh')
```

Let's use groupby to sum the amount spent per year and month.

```
ax = data1.groupby('year_month')['Amount'].sum().sort_index().plot(kind='bar',figsize=(15,6))
```

The following labels the x axis and the y axis.

```
ax.set_xlabel('Month',fontsize=15)
ax.set_ylabel('Amount',fontsize=15)
```

Give a suitable title to the plot.

```
ax.set_title('Revenue Generated for different Months (Dec 2010 - Dec 2011)',fontsize=15)
```

Provide X tick labels.

```
ax.set_xticklabels(('Dec_10','Jan_11','Feb_11','Mar_11','Apr_11','May_11','Jun_11','July_11','Aug_11','Sep_11','Oct_11','Nov_11','Dec_11'),
rotation='horizontal', fontsize=13)=
plt.show()
```

Figure 2-16 shows the output of revenue generated for different months.

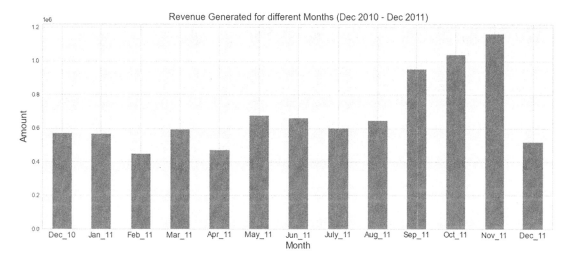

Figure 2-16. *The output*

Item Insights

This segment answers questions like the following.

- Which item was purchased by the greatest number of customers?

- Which is the most sold item based on the sum of sales?

- Which is the most sold item based on the count of orders?

- What are the "first choice" items for the greatest number of invoices?

Most Sold Items Based on Quantity

Create a new pivot table that sums the quantity ordered for each item.

```
most_sold_items_df = data1.pivot_table(index=['StockCode','Descript
ion'], values='Quantity', aggfunc='sum').sort_values(by='Quantity',
ascending=False)
most_sold_items_df.reset_index(inplace=True)
sns.set_style('white')
```

Let's create a bar plot of the ten most ordered items.

```
sns.barplot(y='Description', x='Quantity', data=most_sold_items_
df.head(10))
```

Give a suitable title to the plot.

```
plt.title('Top 10 Items based on No. of Sales', fontsize=14)
plt.ylabel('Item')
```

Figure 2-17 shows the output of the top ten items based on sales.

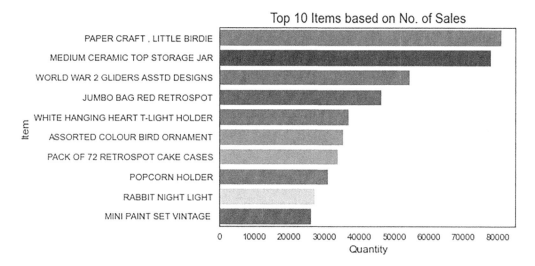

Figure 2-17. *The output*

Items Bought by the Highest Number of Customers

Let's choose WHITE HANGING HEART T-LIGHT HOLDER as an example.

```
product_white_df = data1[data1['Description']=='WHITE HANGING HEART T-LIGHT
HOLDER']
```

```
product_white_df.shape
```

The following is the output.

```
(2028, 13)
```

It denotes that WHITE HANGING HEART T-LIGHT HOLDER has been ordered 2028 times.

```
len(product_white_df.CustomerID.unique())
```

The following is the output.

```
856
```

This means 856 customers ordered WHITE HANGING HEART T-LIGHT HOLDER.

Create a pivot table that displays the sum of unique customers who bought a particular item.

```
most_bought = data1.pivot_table(index=['StockCode','Descripti
on'], values='CustomerID', aggfunc=lambda x: len(x.unique())).sort_
values(by='CustomerID', ascending=False)
most_bought
```

Figure 2-18 shows the output of unique customers who bought a particular item.

StockCode	Description	CustomerID
22423	REGENCY CAKESTAND 3 TIER	881
85123A	WHITE HANGING HEART T-LIGHT HOLDER	856
47566	PARTY BUNTING	708
84879	ASSORTED COLOUR BIRD ORNAMENT	678
22720	SET OF 3 CAKE TINS PANTRY DESIGN	640
...	...	...
21897	POTTING SHED CANDLE CITRONELLA	1
84795C	OCEAN STRIPE HAMMOCK	1
90125E	AMBER BERTIE GLASS BEAD BAG CHARM	1
90128B	BLUE LEAVES AND BEADS PHONE CHARM	1
71143	SILVER BOOK MARK WITH BEADS	1

3897 rows × 1 columns

Figure 2-18. *The output*

Since the WHITE HANGING HEART T-LIGHT HOLDER count matches length 856, the pivot table looks correct for all items.

```
most_bought.reset_index(inplace=True)
sns.set_style('white'
```

Create a bar plot of description (or the item) on the y axis and the sum of unique customers on the x axis.

Plot only the ten most frequently purchased items.

```
sns.barplot(y='Description', x='CustomerID', data=most_bought.head(10))
```

Give a suitable title to the plot.

```
plt.title('Top 10 Items bought by Most no. of Customers', fontsize=14)
plt.ylabel('Item')
```

Figure 2-19 shows the output top ten items by most of the number of customers.

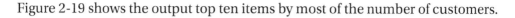

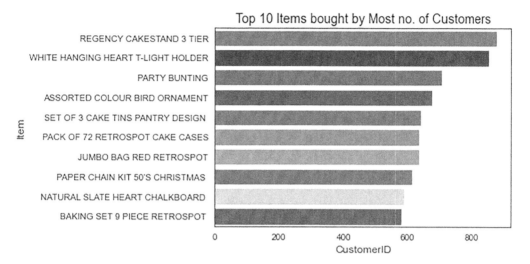

Figure 2-19. *The output*

Most Frequently Ordered Items

Let's prepare data for the word cloud.

```
data1['items'] = data1['Description'].str.replace(' ', '_')
```

Plot the word cloud by using the word cloud library.

```
from wordcloud import WordCloud
plt.rcParams['figure.figsize'] = (20, 20)
wordcloud = WordCloud(background_color = 'white', width = 1200,  height =
1200, max_words = 121).generate(str(data1['items']))
plt.imshow(wordcloud)
plt.axis('off')
plt.title('Most Frequently Bought Items',fontsize = 22)
plt.show()
```

Figure 2-20 shows the word cloud of frequently ordered items.

Figure 2-20. *The output*

Top Ten First Choices

Store all the invoice numbers into a list called l.

```
l = data1['InvoiceNo']
l = l.to_list()
```

The following finds the length of l.

```
len(l)
```

The following is the output.

```
397924
```

Use the set function to find unique invoice numbers only and store them in the invoices list.

```
invoices_list = list(set(l))
```

The following finds the length of the invoices (or the count of unique invoice numbers).

```
len(invoices_list)
```

The following is the output.

```
18536
```

Create an empty list.

```
first_choices_list = []
```

Loop into a list of unique invoice numbers.

```
for i in invoices_list:
    first_purchase_list = data1[data1['InvoiceNo']==i]['items'].reset_
    index(drop=True)[0]

    # Appending
    first_choices_list.append(first_purchase_list)
```

The following creates a first choices list.

```
first_choices_list[:5]
```

The following is the output.

```
['ROCKING_HORSE_GREEN_CHRISTMAS_',
 'POTTERING_MUG',
 'JAM_MAKING_SET_WITH_JARS',
 'TRAVEL_CARD_WALLET_PANTRY',
 'PACK_OF_12_PAISLEY_PARK_TISSUES_']
```

The length of the first choices matches the length of the invoices.

```
len(first_choices_list)
```

The following is the output.

```
18536
```

Use a counter to count repeating first choices.

```
count = Counter(first_choices_list)
```

Store the counter in a DataFrame.

```
df_first_choices = pd.DataFrame.from_dict(count, orient='index').
reset_index()
```

Rename the columns as 'item' and 'count'.

```
df_first_choices.rename(columns={'index':'item', 0:'count'},inplace=True)
```

Sort the DataFrame based on the count.

```
df_first_choices.sort_values(by='count',ascending=False)
```

Figure 2-21 shows the output of the top ten first choices.

	item	count
15	REGENCY_CAKESTAND_3_TIER	203
8	WHITE_HANGING_HEART_T-LIGHT_HOLDER	181
7	RABBIT_NIGHT_LIGHT	155
118	PARTY_BUNTING	122
28	Manual	119
...	...	...
2041	CAKE_SHOP__STICKER_SHEET	1
538	PINK_POLKADOT_KIDS_BAG	1
2045	RIBBON_REEL_SOCKS_AND_MITTENS	1
2046	DOG_TOY_WITH_PINK_CROCHET_SKIRT	1
2634	ELEPHANT_BIRTHDAY_CARD_	1

2635 rows × 2 columns

Figure 2-21. *The output*

```
plt.subplots(figsize=(20,10))
sns.set_style('white')
```

Let's create a bar plot.

```
sns.barplot(y='item', x='count', data=df_first_choices.sort_values(by='count',
ascending=False).head(10))
```

Give a suitable title to the plot.

```
plt.title('Top 10 First Choices', fontsize=14)
plt.ylabel('Item')
```

Figure 2-22 shows the output of the top ten first choices.

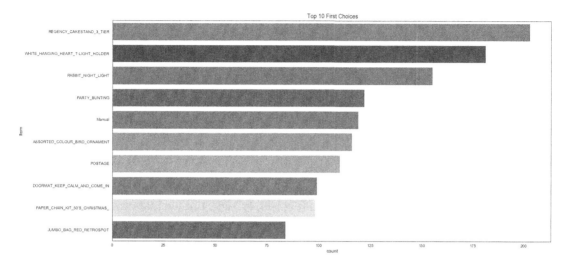

Figure 2-22. *The output*

Frequently Bought Together (MBA)

This segment answers questions like the following.

- Which items are frequently bought together?

- If a user buys an item X, which item is he/she likely to buy next?

Let's use group by function to create a market basket DataFrame, which specifies if an item is present in a particular invoice number for all items and all invoices.

The following denotes the quantity in the invoice number, which must be fixed.

```
market_basket = (data1.groupby(['InvoiceNo', 'Description'])['Quantity'].
sum().unstack().reset_index().fillna(0).set_index('InvoiceNo'))
market_basket.head(10)
```

Figure 2-23 shows the output of total quantity, grouped by invoice and description.

Description	4 PURPLE FLOCK DINNER CANDLES	50'S CHRISTMAS GIFT BAG LARGE	DOLLY GIRL BEAKER	I LOVE LONDON MINI BACKPACK	I LOVE LONDON MINI RUCKSACK	NINE DRAWER OFFICE TIDY	OVAL WALL MIRROR DIAMANTE	RED SPOT GIFT BAG LARGE	SET 2 TEA TOWELS I LOVE LONDON	SPACEBOY BABY GIFT SET	...	ZINC STAR T-LIGHT HOLDER	ZINC SWEETHEART SOAP DISH	
InvoiceNo														
536365	0.0	0.0	0.0	0.0	0.0	0.0	0.0	0.0	0.0	0.0	...	0.0	0.0	
536366	0.0	0.0	0.0	0.0	0.0	0.0	0.0	0.0	0.0	0.0	...	0.0	0.0	
536367	0.0	0.0	0.0	0.0	0.0	0.0	0.0	0.0	0.0	0.0	...	0.0	0.0	
536368	0.0	0.0	0.0	0.0	0.0	0.0	0.0	0.0	0.0	0.0	...	0.0	0.0	
536369	0.0	0.0	0.0	0.0	0.0	0.0	0.0	0.0	0.0	0.0	...	0.0	0.0	
536370	0.0	0.0	0.0	0.0	0.0	0.0	0.0	0.0	24.0	0.0	...	0.0	0.0	
536371	0.0	0.0	0.0	0.0	0.0	0.0	0.0	0.0	0.0	0.0	...	0.0	0.0	
536372	0.0	0.0	0.0	0.0	0.0	0.0	0.0	0.0	0.0	0.0	...	0.0	0.0	
536373	0.0	0.0	0.0	0.0	0.0	0.0	0.0	0.0	0.0	0.0	...	0.0	0.0	
536374	0.0	0.0	0.0	0.0	0.0	0.0	0.0	0.0	0.0	0.0	...	0.0	0.0	

10 rows × 3877 columns

Figure 2-23. *The output*

This output gets the quantity ordered (e.g., 48,24,126), but we just want to know if an item was purchased or not.

So, let's encode the units as 1 (if purchased) or 0 (not purchased).

```
def encode_units(x):
    if x < 1:
        return 0
    if x >= 1:
        return 1
market_basket = market_basket.applymap(encode_units)
market_basket.head(10)
```

Description	4 PURPLE FLOCK DINNER CANDLES	50'S CHRISTMAS GIFT BAG LARGE	DOLLY GIRL BEAKER	I LOVE LONDON MINI BACKPACK	I LOVE LONDON MINI RUCKSACK	NINE DRAWER OFFICE TIDY	OVAL WALL MIRROR DIAMANTE	RED SPOT GIFT BAG LARGE	SET 2 TEA TOWELS I LOVE LONDON	SPACEBOY BABY GIFT SET	...	ZINC STAR T-LIGHT HOLDER	ZINC SWEETHEART SOAP DISH	
InvoiceNo														
536365	0	0	0	0	0	0	0	0	0	0	...	0	0	
536366	0	0	0	0	0	0	0	0	0	0	...	0	0	
536367	0	0	0	0	0	0	0	0	0	0	...	0	0	
536368	0	0	0	0	0	0	0	0	0	0	...	0	0	
536369	0	0	0	0	0	0	0	0	0	0	...	0	0	
536370	0	0	0	0	0	0	0	0	1	0	...	0	0	
536371	0	0	0	0	0	0	0	0	0	0	...	0	0	
536372	0	0	0	0	0	0	0	0	0	0	...	0	0	
536373	0	0	0	0	0	0	0	0	0	0	...	0	0	
536374	0	0	0	0	0	0	0	0	0	0	...	0	0	

10 rows × 3877 columns

Figure 2-24. *The output*

Apriori Algorithm Concepts

Refer to Chapter 1 for more information.

Figure 2-25 explains the support.

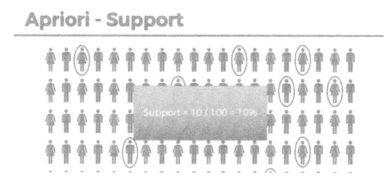

Figure 2-25. *Support*

Let's look at an example. If 10 out of 100 users purchase milk, support for milk is 10/100 = 10%. The calculation formula is shown in Figure 2-26.

$$\text{Movie Recommendation:} \quad \text{support}(M) = \frac{\#\ \text{user watchlists containing } M}{\#\ \text{user watchlists}}$$

$$\text{Market Basket Optimisation:} \quad \text{support}(I) = \frac{\#\ \text{transactions containing } I}{\#\ \text{transactions}}$$

Figure 2-26. *Formula*

Suppose you are looking to build a relationship between milk and bread. If 7 out of 40 milk buyers also buy bread, then confidence = 7/40 = 17.5%

Figure 2-27 explains confidence.

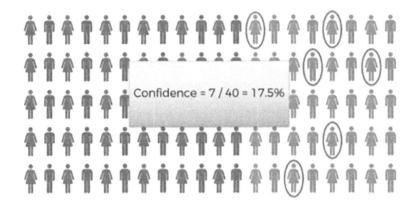

Figure 2-27. *Confidence*

The formula to calculate confidence is shown in Figure 2-28.

Movie Recommendation: $confidence(M_1 \to M_2) = \dfrac{\text{\# user watchlists containing } M_1 \text{ and } M_2}{\text{\# user watchlists containing } M_1}$

Market Basket Optimisation: $confidence(I_1 \to I_2) = \dfrac{\text{\# transactions containing } I_1 \text{ and } I_2}{\text{\# transactions containing } I_1}$

Figure 2-28. *Formula*

The basic formula is lift = confidence/support.

So here, lift = 17.5/10 = 1.75.

Figure 2-29 explains lift and the formula.

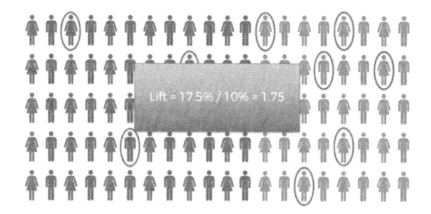

Movie Recommendation:

$$\text{lift}(M_1 \rightarrow M_2) = \frac{\text{confidence}(M_1 \rightarrow M_2)}{\text{support}(M_2)}$$

Market Basket Optimisation:

$$\text{lift}(I_1 \rightarrow I_2) = \frac{\text{confidence}(I_1 \rightarrow I_2)}{\text{support}(I_2)}$$

Figure 2-29. *Lift*

Association Rules

Association rule mining finds interesting associations and relationships among large sets of data items. This rule shows how frequently an item set occurs in a transaction. A market basket analysis is performed based on the rules created from the dataset.

Figure 2-30 explains the association rule.

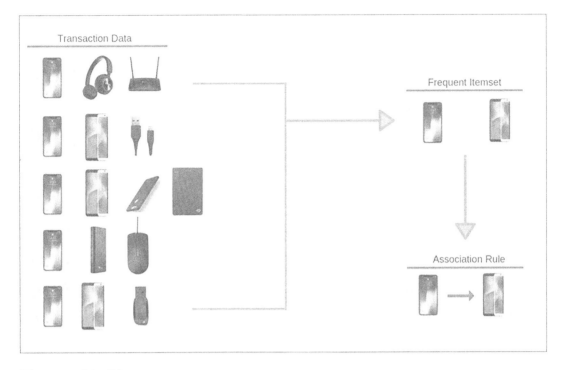

Figure 2-30.   *The output*

Figure 2-30 shows that out of the five transactions in which a mobile phone was purchased, three included a mobile screen guard. Thus, it should be recommended.

Implementation Using mlxtend

Let's look at a sample item.

```
product_wooden_star_df = market_basket.loc[market_basket['WOODEN STAR
CHRISTMAS SCANDINAVIAN']==1]
```

If A => then B

Use the apriori algorithm and create association rules for the sample item.

Apply the apriori algorithm to product_wooden_star_df.

```
itemsets_frequent = apriori(product_wooden_star_df, min_support=0.15, use_
colnames=True)
```

Store the association rules into rules.

```
prod_wooden_star_rules = association_rules(itemsets_frequent,
metric="lift", min_threshold=1)
```

Sort the rules on lift and support.

```
prod_wooden_star_rules.sort_values(['lift','support'],ascending=False).
reset_index(drop=True).head()
```

Figure 2-31 shows the output of apriori algorithm.

	antecedents	consequents	antecedent support	consequent support	support	confidence	lift	leverage	conviction
0	(WOODEN HEART CHRISTMAS SCANDINAVIAN)	(WOODEN TREE CHRISTMAS SCANDINAVIAN)	0.736721	0.521940	0.420323	0.570533	1.093101	0.035799	1.113147
1	(WOODEN TREE CHRISTMAS SCANDINAVIAN)	(WOODEN HEART CHRISTMAS SCANDINAVIAN)	0.521940	0.736721	0.420323	0.805310	1.093101	0.035799	1.352299
2	(WOODEN HEART CHRISTMAS SCANDINAVIAN, WOODEN S...	(WOODEN TREE CHRISTMAS SCANDINAVIAN)	0.736721	0.521940	0.420323	0.570533	1.093101	0.035799	1.113147
3	(WOODEN STAR CHRISTMAS SCANDINAVIAN, WOODEN TR...	(WOODEN HEART CHRISTMAS SCANDINAVIAN)	0.521940	0.736721	0.420323	0.805310	1.093101	0.035799	1.352299
4	(WOODEN HEART CHRISTMAS SCANDINAVIAN)	(WOODEN STAR CHRISTMAS SCANDINAVIAN, WOODEN TR...	0.736721	0.521940	0.420323	0.570533	1.093101	0.035799	1.113147

Figure 2-31. *The output*

Creating a Function

Create a new function to pass an item name. It returns the items that are bought together frequently. In other words, it returns the items that are likely to be bought by the user because they bought the item passed into the function.

```
def bought_together_frequently(item):

    # df of item passed
    df_item = market_basket.loc[market_basket[item]==1]

    # Apriori algorithm
    itemsets_frequent = apriori(df_item, min_support=0.15, use_
    colnames=True)

    # Storing association rules
    a_rules = association_rules(itemsets_frequent, metric="lift", min_
    threshold=1)

    # Sorting on lift and support
```

```
a_rules.sort_values(['lift','support'],ascending=False).reset_
index(drop=True)

print('Items frequently bought together with {0}'.format(item))

# Returning top 6 items with highest lift and support
return a_rules['consequents'].unique()[:6]
```

Example 1 is as follows.

```
bought_together_frequently('WOODEN STAR CHRISTMAS SCANDINAVIAN')
```

The following is the output.

```
Items frequently bought together with WOODEN STAR CHRISTMAS SCANDINAVIAN
array([frozenset({"PAPER CHAIN KIT 50'S CHRISTMAS "}),
       frozenset({'WOODEN HEART CHRISTMAS SCANDINAVIAN'}),
       frozenset({'WOODEN STAR CHRISTMAS SCANDINAVIAN'}),
       frozenset({'SET OF 3 WOODEN HEART DECORATIONS'}),
       frozenset({'SET OF 3 WOODEN SLEIGH DECORATIONS'}),
       frozenset({'SET OF 3 WOODEN STOCKING DECORATION'})], dtype=object)
```

Example 2 is as follows.

```
bought_together_frequently('WHITE METAL LANTERN')
```

The following is the output.

```
Items frequently bought together with WHITE METAL LANTERN
array([frozenset({'LANTERN CREAM GAZEBO '}),
       frozenset({'WHITE METAL LANTERN'}),
       frozenset({'REGENCY CAKESTAND 3 TIER'}),
       frozenset({'WHITE HANGING HEART T-LIGHT HOLDER'})], dtype=object)
```

Example 3 is as follows.

```
bought_together_frequently('JAM MAKING SET WITH JARS')
```

The following is the output.

```
Items frequently bought together with JAM MAKING SET WITH JARS
array([frozenset({'JAM MAKING SET WITH JARS'}),
       frozenset({'JAM MAKING SET PRINTED'}),
```

```
frozenset({'PACK OF 72 RETROSPOT CAKE CASES'}),
frozenset({'RECIPE BOX PANTRY YELLOW DESIGN'}),
frozenset({'REGENCY CAKESTAND 3 TIER'}),
frozenset({'SET OF 3 CAKE TINS PANTRY DESIGN '})], dtype=object)
```

Validation

JAM MAKING SET PRINTED is a part of invoice 536390, so let's print all the items from this invoice and cross-check it.

```
data1[data1 ['InvoiceNo']=='536390']
```

Figure 2-32 shows the output of filtered data.

ckCode	year_month	month	day	hour	Description	Quantity	InvoiceDate	UnitPrice	CustomerID	Country	Amount	items
22941	201012	12	3	10	CHRISTMAS LIGHTS 10 REINDEER	2	2010-12-01 10:19:00	8.50	17511.0	United Kingdom	17.00	CHRISTMAS_LIGHTS_10_REINDEER
22960	201012	12	3	10	JAM MAKING SET WITH JARS	12	2010-12-01 10:19:00	3.75	17511.0	United Kingdom	45.00	JAM_MAKING_SET_WITH_JARS
22961	201012	12	3	10	JAM MAKING SET PRINTED	12	2010-12-01 10:19:00	1.45	17511.0	United Kingdom	17.40	JAM_MAKING_SET_PRINTED
22962	201012	12	3	10	JAM JAR WITH PINK LID	48	2010-12-01 10:19:00	0.72	17511.0	United Kingdom	34.56	JAM_JAR_WITH_PINK_LID
22963	201012	12	3	10	JAM JAR WITH GREEN LID	48	2010-12-01 10:19:00	0.72	17511.0	United Kingdom	34.56	JAM_JAR_WITH_GREEN_LID
22968	201012	12	3	10	ROSE COTTAGE KEEPSAKE BOX	8	2010-12-01 10:19:00	8.50	17511.0	United Kingdom	68.00	ROSE_COTTAGE_KEEPSAKE_BOX_
84970S	201012	12	3	10	HANGING HEART ZINC T-LIGHT HOLDER	144	2010-12-01 10:19:00	0.64	17511.0	United Kingdom	92.16	HANGING_HEART_ZINC_T-LIGHT_HOLDER
22910	201012	12	3	10	PAPER CHAIN KIT VINTAGE CHRISTMAS	40	2010-12-01 10:19:00	2.55	17511.0	United Kingdom	102.00	PAPER_CHAIN_KIT_VINTAGE_CHRISTMAS
20668	201012	12	3	10	DISCO BALL CHRISTMAS DECORATION	288	2010-12-01 10:19:00	0.10	17511.0	United Kingdom	28.80	DISCO_BALL_CHRISTMAS_DECORATION
85123A	201012	12	3	10	WHITE HANGING HEART T-LIGHT HOLDER	64	2010-12-01 10:19:00	2.55	17511.0	United Kingdom	163.20	WHITE_HANGING_HEART_T-LIGHT_HOLDER

Figure 2-32.   *The output*

There are some common items between the recommendations from the bought_together_frequently function and the invoice.

Thus, the recommender is performing well.

Visualization of Association Rules

Let's try visualization techniques on the WOODEN STAR DataFrame used earlier.

```
support=prod_wooden_star_rules.support.values
confidence=prod_wooden_star_rules.confidence.values
```

The following creates a scatter plot.

```
import networkx as nx
import random
import matplotlib.pyplot as plt

for i in range (len(support)):
    support[i] = support[i] + 0.0025 * (random.randint(1,10) - 5)
    confidence[i] = confidence[i] + 0.0025 * (random.randint(1,10) - 5)

# Creating a scatter plot of support v confidence
plt.scatter(support, confidence,    alpha=0.5, marker="*")
plt.xlabel('support')
plt.ylabel('confidence')
plt.show()
```

Figure 2-33 shows the confidence vs. support.

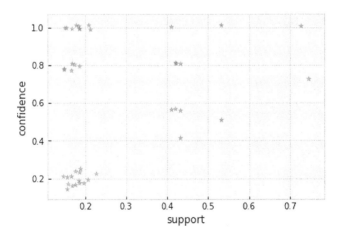

Figure 2-33. *The output*

Let's plot a graphical representation.

```python
def graphing_wooden_star(wooden_star_rules, no_of_rules):
    Graph1 = nx.DiGraph()
    color_map=[]
    N = 50
    colors = np.random.rand(N)
    strs=['R0', 'R1', 'R2', 'R3', 'R4', 'R5', 'R6', 'R7', 'R8', 'R9',
    'R10', 'R11']

    for i in range (no_of_rules):

        # adding as many nodes as number of rules requested by user
        Graph1.add_nodes_from(["R"+str(i)])

    # adding antecedents to the nodes
    for a in wooden_star_rules.iloc[i]['antecedents']:

        Graph1.add_nodes_from([a])

        Graph1.add_edge(a, "R"+str(i), color=colors[i] , weight = 2)

    # adding consequents to the nodes
    for c in wooden_star_rules.iloc[i]['consequents']:

            Graph1.add_nodes_from([c])

            Graph1.add_edge("R"+str(i), c, color=colors[i],  weight=2)

    for node in Graph1:
        found_a_string = False
        for item in strs:
            if node==item:
                found_a_string = True
        if found_a_string:
            color_map.append('yellow')
        else:
            color_map.append('green')

    edges = Graph1.edges()
    colors = [Graph1[u][v]['color'] for u,v in edges]
    weights = [Graph1[u][v]['weight'] for u,v in edges]
```

```
pos = nx.spring_layout(Graph1, k=16, scale=1)
nx.draw(Graph1, pos, edges=edges, node_color = color_map, edge_
color=colors, width=weights, font_size=16, with_labels=False)

for p in pos:  # raise text positions
        pos[p][1] += 0.07
nx.draw_networkx_labels(G1, pos)
plt.show()
```

Figure 2-34 shows the graphical representation.

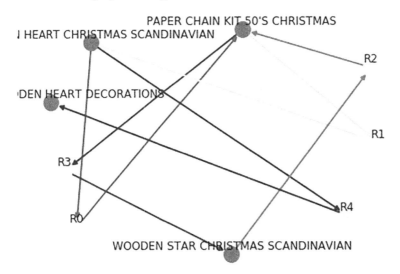

Figure 2-34. *The output*

```
def visualize_rules(item, no_of_rules):

    # df of item passed
    df_item = market_basket.loc[market_basket[item]==1]

    # Apriori algorithm
    itemsets_frequent = apriori(df_item, min_support=0.15, use_
    colnames=True)

    # Storing association rules
    a_rules = association_rules(itemsets_frequent, metric="lift", min_
    threshold=1)
```

```
# Sorting on lift and support
a_rules.sort_values(['lift','support'],ascending=False).reset_
index(drop=True)
print('Items frequently bought together with {0}'.format(item))

# Returning top 6 items with highest lift and support
print(a_rules['consequents'].unique()[:6])

support = a_rules.support.values
confidence = a_rules.confidence.values

for i in range (len(support)):
    support[i] = support[i] + 0.0025 * (random.randint(1,10) - 5)
    confidence[i] = confidence[i] + 0.0025 * (random.randint(1,10) - 5)

# Creating scatter plot of support v confidence
plt.scatter(support, confidence, alpha=0.5, marker="*")
plt.title('Support vs Confidence graph')
plt.xlabel('support')
plt.ylabel('confidence')
plt.show()

# Creating a new digraph
Graph2 = nx.DiGraph()

color_map=[]
N = 50
colors = np.random.rand(N)
strs=['R0', 'R1', 'R2', 'R3', 'R4', 'R5', 'R6', 'R7', 'R8', 'R9',
'R10', 'R11']

# adding as many nodes as number of rules requested by user
for i in range (no_of_rules):
    Graph2.add_nodes_from(["R"+str(i)])

# adding antecedents to the nodes
for a in a_rules.iloc[i]['antecedents']:

    Graph2.add_nodes_from([a])
```

```
        Graph2.add_edge(a, "R"+str(i), color=colors[i] , weight = 2)

    # adding consequents to the nodes
    for c in a_rules.iloc[i]['consequents']:
        Graph2.add_nodes_from([c])
        Graph2.add_edge("R"+str(i), c, color=colors[i],  weight=2)

    for node in Graph2:
        found_a_string = False
        for item in strs:
            if node==item:
                found_a_string = True
        if found_a_string:
            color_map.append('yellow')
        else:
            color_map.append('green')

    print('Visualization of Rules:')

    edges = Graph2.edges()
    colors = [Graph2[u][v]['color'] for u,v in edges]
    weights = [Graph2[u][v]['weight'] for u,v in edges]

    pos = nx.spring_layout(Graph2, k=16, scale=1)
    nx.draw(Graph2, pos, edges=edges, node_color = color_map, edge_
    color=colors, width=weights, font_size=16, with_labels=False)

    for p in pos:  # raise text positions
        pos[p][1] += 0.07
    nx.draw_networkx_labels(Graph2, pos)
    plt.show()
```

Example 1 is as follows.

```
visualize_rules('WOODEN STAR CHRISTMAS SCANDINAVIAN',4)
```

Figure 2-35 shows items frequently bought along with WOODEN STAR CHRISTMAS SCANDINAVIAN.

Figure 2-35. *The output*

Figure 2-36 shows the visualization of rules.

```
[frozenset({'WOODEN HEART CHRISTMAS SCANDINAVIAN'})
 frozenset({"PAPER CHAIN KIT 50'S CHRISTMAS "})
 frozenset({'WOODEN STAR CHRISTMAS SCANDINAVIAN'})
 frozenset({'SET OF 3 WOODEN HEART DECORATIONS'})
 frozenset({'SET OF 3 WOODEN SLEIGH DECORATIONS'})
 frozenset({'SET OF 3 WOODEN STOCKING DECORATION'})]
```

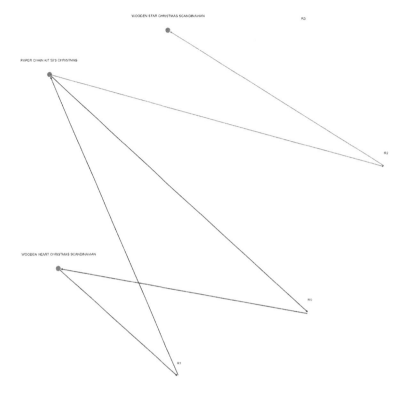

Figure 2-36. *The output*

Example 2 is as follows.

```
visualize_rules('JAM MAKING SET WITH JARS',6)
```

Figure 2-37 shows the items frequently bought together with JAM MAKING SET WITH JARS.

Figure 2-37. *The output*

Figure 2-38 shows the visualization of rules.

```
[frozenset({'JAM MAKING SET WITH JARS'})
 frozenset({'JAM MAKING SET PRINTED'})
 frozenset({'PACK OF 72 RETROSPOT CAKE CASES'})
 frozenset({'RECIPE BOX PANTRY YELLOW DESIGN'})
 frozenset({'REGENCY CAKESTAND 3 TIER'})
 frozenset({'SET OF 3 CAKE TINS PANTRY DESIGN '})]
```

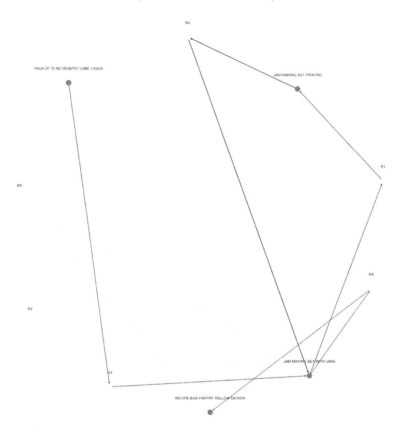

Figure 2-38. *The output*

Summary

In this chapter, you learned how to build a recommendation system based on market basket analysis. You also learned how to fetch items that are frequently purchased together and offer suggestions to users. Most e-commerce sites use this method to showcase items bought together. This chapter implemented this method in Python using an e-commerce example.

CHAPTER 3

Content-Based Recommender Systems

Content-based filtering is used to recommend products or items very similar to those being clicked or liked. User recommendations are based on the description of an item and a profile of the user's interests. Content-based recommender systems are widely used in e-commerce platforms. It is one of the basic algorithms in a recommendation engine. Content-based filtering can be triggered for any event; for example, on click, on purchase, or add to cart.

If you use any e-commerce platform, for example, Amazon.com, a product's page shows recommendations in the "related products" section. How to generate these recommendations is discussed in this chapter.

Approach

The following steps build a content-based recommender engine.

1. Do the data collection (should have complete item description).

2. Do the data preprocessing.

3. Convert text to features.

4. Perform similarity measures.

5. Recommend products.

Figure 3-1 illustrates these steps.

© Akshay Kulkarni, Adarsha Shivananda, Anoosh Kulkarni, V Adithya Krishnan 2023
A. Kulkarni et al., *Applied Recommender Systems with Python*, https://doi.org/10.1007/978-1-4842-8954-9_3

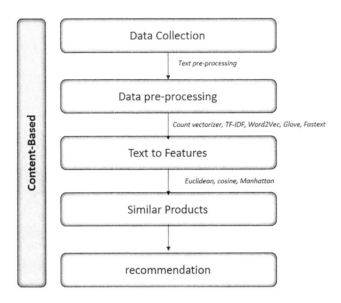

Figure 3-1. *Steps*

Implementation

The following installs and imports the required libraries.

```
#Importing the libraries

import pandas as pd
from sklearn.feature_extraction.text import CountVectorizer
from sklearn.metrics.pairwise import cosine_similarity, manhattan_
distances, euclidean_distances
from sklearn.feature_extraction.text import TfidfVectorizer
import re
from gensim import models
import numpy as np
import matplotlib.pyplot as plt
import matplotlib.style
%matplotlib inline
from gensim.models import FastText as ft
from IPython.display import Image
import os
```

Data Collection and Downloading Word Embeddings

Let's look at an e-commerce dataset. Download the dataset from GitHub.

You can download the required pre-trained models from the following URLs.

- Word2vec: `https://drive.google.com/uc?id=0B7XkCwpI5KDYNlNUT TlSS21pQmM`

- GloVe: `https://nlp.stanford.edu/data/glove.6B.zip`

- fastText: `https://dl.fbaipublicfiles.com/fasttext/vectors-crawl/cc.en.300.bin.gz`

Importing the Data as a DataFrame (pandas)

The following imports the data.

```
Content_df = pd.read_csv("Rec_sys_content.csv")
```

The following prints the top five rows of the DataFrame.

```
#Viewing Top 5 Rows
Content_df.head(5)
```

Figure 3-2 shows the output of the first five rows.

	StockCode	Product Name	Description	Category	Brand	Unit Price
0	22629	Ganma Superheroes Ordinary Life Case For Samsu...	New unique design, great gift.High quality pla...	Cell Phones|Cellphone Accessories|Cases & Prot...	Ganma	13.99
1	21238	Eye Buy Express Prescription Glasses Mens Wome...	Rounded rectangular cat-eye reading glasses. T...	Health|Home Health Care|Daily Living Aids	Eye Buy Express	19.22
2	22181	MightySkins Skin Decal Wrap Compatible with Ni...	Each Nintendo 2DS kit is printed with super-hi...	Video Games|Video Game Accessories|Accessories ...	Mightyskins	14.99
3	84879	Mediven Sheer and Soft 15-20 mmHg Thigh w/ Lac...	The sheerest compression stocking in its class...	Health|Medicine Cabinet|Braces & Supports	Medi	62.38
4	84836	Stupell Industries Chevron Initial Wall D cor	Features: -Made in the USA. - Sawtooth hanger o...	Home Improvement|Paint|Wall Decals|All Wall De...	Stupell Industries	35.99

Figure 3-2.   *The output*

Let's check the internal structure of each column in the dataset.

```
# Data Info
Content_df.info()
```

The following is the output.

```
<class 'pandas.core.frame.DataFrame'>
RangeIndex: 3958 entries, 0 to 3957
Data columns (total 6 columns):
 #   Column         Non-Null Count   Dtype
---  ------         --------------   -----
 0   StockCode      3958 non-null    object
 1   Product Name   3958 non-null    object
 2   Description    3958 non-null    object
 3   Category       3856 non-null    object
 4   Brand          3818 non-null    object
 5   Unit Price     3943 non-null    float64
dtypes: float64(1), object(5)
memory usage: 185.7+ KB
```

Preprocessing the Data

Before cleaning the data, check the number of rows and columns and then check for null values.

```
Content_df.shape
```

The following is the output.

```
(3958, 6)
```

```
# Total Null Values in Data
Content_df.isnull().sum(axis = 0)
```

The following is the output.

```
StockCode          0
Product Name       0
Description        0
Category         102
Brand            140
Unit Price        15
dtype: int64
```

There are a few null values present in the dataset. However, let's focus on Product Name and Description to build a content-based recommendation engine. Removing nulls from Category, Brand, and Unit Price is not required.

Now, let's load the pre-trained models.

```
#Importing Word2Vec
word2vecModel = models.KeyedVectors.load_word2vec_format('GoogleNews-
vectors-negative300.bin.gz', binary=True)
```

```
#Importing FastText
fasttext_model=ft.load_fasttext_format("cc.en.300.bin.gz")
```

```
#Import Glove
glove_df = pd.read_csv('glove.6B.300d.txt', sep=" ",
                       quoting=3, header=None, index_col=0)
glove_model = {key: value.values for key, value in glove_df.T.items()}
```

As discussed, the Product Name and Description columns of text data are used to build a content-based recommendation engine. Text preprocessing is mandatory. It is followed by text-to-feature conversion.

The following describes the preprocessing steps.

1. Remove duplicates.

2. Convert the string to lowercase.

3. Remove special characters.

```
## Combining Product and Description
Content_df['Description'] = Content_df['Product Name'] + ' ' +Content_
df['Description']
```

```
# Dropping Duplicates and keeping first record
unique_df = Content_df.drop_duplicates(subset=['Description'],
keep='first')
```

```
# Converting String to Lower Case
unique_df['desc_lowered'] = unique_df['Description'].apply(lambda x:
x.lower())
```

```
# Remove Stop special Characters
```

```
unique_df['desc_lowered'] = unique_df['desc_lowered'].apply(lambda x:
re.sub(r'[^\w\s]', '', x))

# Coverting Description to List
desc_list = list(unique_df['desc_lowered'])

unique_df= unique_df.reset_index(drop=True)
```

Text to Features

Once the text preprocessing is done, let's focus on converting the preprocessed text into features.

There are several methods to convert text to features.

- One-hot encoding (OHE)

- CountVectorizer

- TF-IDF

The following are word embedding tools.

- Word2vec

- fastText

- GloVe

Since machines or algorithms cannot understand the text, a key task in natural language processing (NLP) is converting text data into numerical data called *features*. There are different techniques to do this. Let's discuss them briefly.

One-Hot Encoding (OHE)

OHE is the basic and simple way of converting text to numbers or features. It converts all the tokens in the corpus into columns, as shown in Table 3-1. After that, against every observation, it tags as 1 if the word is present; otherwise, 0.

Table 3-1. *OHE*

	One	**Hot**	**Encoding**
One Hot	1	1	0
Hot	0	1	0
Encoding	0	0	1

CountVectorizer

The drawback to the OHE approach is if a word appears multiple times in a sentence, it gets the same importance as any other word that appears only once. CountVectorizer helps overcome this because it counts the tokens present in an observation instead of tagging everything as 1 or 0.

Table 3-2 demonstrates CountVectorizer.

Table 3-2. *CountVectorizer*

	AI	**new**	**Learn**	**.......**
AI is new. AI is everywhere.	2	1	0	....
Learn AI. Learn NLP.	1	1	2	....
NLP is cool.	0	0	0	....

Term Frequency–Inverse Document Frequency (TF-IDF)

CountVectorizer won't answer all questions. If the length of the sentences is inconsistent or a word is repeated in all the sentences, it becomes tricky. TF-IDF addresses these problems.

The *term frequency* (TF) is the "number of times the token appeared in a corpus doc divided by the total number of tokens."

The *inverse document frequency* (IDF) is a log of the total number of such corpus docs in overall docs we have divided by the number of overall docs with the selected word. It helps provide more weight to rare words in the corpus.

Multiplying them gives the TF-IDF vector for a word in the corpus.

$$tfidf_{i,j} = tf_{i,j} \cdot idf_i$$

$$tf_{i,j} = \frac{Number\ of\ times\ term\ i\ appears\ in\ document\ j}{Total\ number\ of\ terms\ in\ document\ j}$$

$$idf_i = log\left(\frac{Total\ number\ of\ documents}{Number\ of\ documents\ with\ term\ i\ in\ it}\right)$$

Word Embeddings

Even though TF-IDF is widely used, it doesn't capture the context of a word or sentence. Word embeddings solve this problem. Word embeddings help capture context and the semantic and syntactic similarity between the words. They generate a vector that captures the context and semantics using shallow neural networks.

In recent years, there have been many advancements in this field, including the following tools.

- Word2vec

- GloVe

- fastText

- Elmo

- SentenceBERT

- GPT

For more information about these concepts, please refer to our second edition book on NLP, *Natural Language Processing Recipes: Unlocking Text Data with Machine Learning and Deep Learning Using Python* (Apress, 2021).

The pre-trained models (word embeddings)—GloVe, Word2vec, and fastText—have been imported/loaded. Now let's import the CountVectorizer and TF-IDF.

```
#Importing Count Vectorizer
cnt_vec = CountVectorizer(stop_words='english')
```

```
# Importing IFIDF
tfidf_vec = TfidfVectorizer(stop_words='english', analyzer='word', ngram_
range=(1,3))
```

Similarity Measures

Once text is converted to features, the next step is to build a content-based model. The similarity measures must get similar vectors.

There are three types of similarity measures.

- Euclidean distance

- Cosine similarity

- Manhattan distance

Note   We have not yet converted text to features; we only loaded all the methods. They are used later.

Euclidean Distance

Euclidean distance is calculated by taking the sum of the squared differences between two vectors and then applying the square root.

Figure 3-3 explains the Euclidean distance.

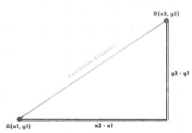

$$Euclidean\,(A, B) = \sqrt{(x_2 - x_1)^2 + (y_2 - y_1)^2}$$

Figure 3-3. *Euclidean distance*

Cosine Similarity

Cosine similarity is the cosine of the angle between two n-dimensional vectors in an N-dimensional space. It is the dot product of the two vectors divided by the product of the two vectors' lengths (or magnitudes).

Figure 3-4 explains cosine similarity.

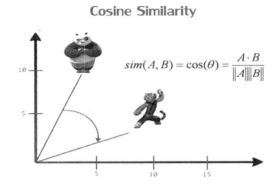

Figure 3-4. *Cosine similarity*

Manhattan Distance

The Manhattan distance is the sum of the absolute differences between two vectors.

Figure 3-5 explains the Manhattan distance.

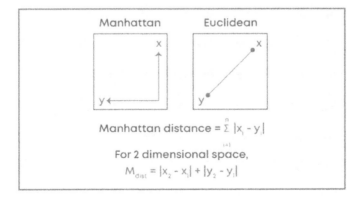

Figure 3-5. *Manhattan distance*

Let's write functions for all three types of similarity measures.

```
#Euclidean distance
```

```python
def find_euclidean_distances(sim_matrix, index, n=10):

    # Getting Score and Index
    result = list(enumerate(sim_matrix[index]))

    # Sorting the Score and taking top 10 products
    sorted_result = sorted(result,key=lambda x:x[1],reverse=False)[1:10+1]

    # Mapping index with data
    similar_products =  [{'value': unique_df.iloc[x[0]]['Product Name'],
    'score' : round(x[1], 2)} for x in sorted_result]

    return similar_products

#Cosine similarity
def find_similarity(cosine_sim_matrix, index, n=10):

    # calculate cosine similarity between each vectors
    result = list(enumerate(cosine_sim_matrix[index]))

    # Sorting the Score
    sorted_result = sorted(result,key=lambda x:x[1],reverse=True)[1:n+1]

    similar_products =  [{'value': unique_df.iloc[x[0]]['Product Name'],
    'score' : round(x[1], 2)} for x in sorted_result]

    return similar_products

#Manhattan distance
def find_manhattan_distance(sim_matrix, index, n=10):

    # Getting Score and Index
    result = list(enumerate(sim_matrix[index]))

    # Sorting the Score and taking top 10 products
    sorted_result = sorted(result,key=lambda x:x[1],reverse=False)[1:10+1]

    # Mapping index with data
    similar_products =  [{'value': unique_df.iloc[x[0]]['Product Name'],
    'score' : round(x[1], 2)} for x in sorted_result]

    return similar_products
```

Build a Model Using CountVectorizer

Using CountVectorizer features, let's write a function that recommends the top ten most similar products.

```
#Comparing similarity to get the top matches using count Vec

def get_recommendation_cv(product_id, df, similarity, n=10):

    row = df.loc[df['Product Name'] == product_id]
    index = list(row.index)[0]
    description = row['desc_lowered'].loc[index]

    #Create vector using Count Vectorizer

    count_vector = cnt_vec.fit_transform(desc_list)

    if similarity == "cosine":
        sim_matrix = cosine_similarity(count_vector)
        products = find_similarity(sim_matrix , index)

    elif similarity == "manhattan":
        sim_matrix = manhattan_distances(count_vector)
        products = find_manhattan_distance(sim_matrix , index)

    else:
        sim_matrix = euclidean_distances(count_vector)
        products = find_euclidean_distances(sim_matrix , index)

    return products
```

The following is the input for this function.

- Product id = Mentions the product name/description for which you need similar items

- df = Passes preprocessed data

- similarity = Mentions which similarity method must run

- n = Number of recommendations

Now let's get similar recommendations for one product.

The following is an example.

```
product_id = 'Vickerman 14" Finial Drop Christmas Ornaments, Pack of 2'
```

Next, get recommendations using cosine similarity for CountVectorizer features.

```
# Cosine Similarity
get_recommendation_cv(product_id, unique_df, similarity = "cosine", n=10)
```

Figure 3-6 shows the output of cosine similarity for CountVectorizer features.

```
[{'value': 'Fancyleo Christmas Glasses Frames 2 Pack Glittered Eyeglasses Glasses Set No Lens Kids Family Xmas Party Ornamen
ts Gift',
  'score': 0.28},
 {'value': 'storefront christmas LED Decoration Light Gold Color Star Shape Vine Wedding Party event',
  'score': 0.2},
 {'value': '16 inches 40 inches "MERRY CHRISTMAS" Letter Foil Inflated Balloon Float Helium Aluminum Balloons for Christmas
Decoration',
  'score': 0.19},
 {'value': '12ct Light Gunmetal Gray Shatterproof 4-Finish Christmas Ball Ornaments 4" (100mm)',
  'score': 0.19},
 {'value': 'PeanutsÃ‚Â Valentine Sign Craft Kit (Pack of 12)', 'score': 0.13},
 {'value': 'Simplicity 3 Pack Elastic Hair Ties White/Pink/Pink Leopard, 24 Count',
  'score': 0.13},
 {'value': '3 Pack Newbee Fashion- "Butterfly" Thin Design Gold Emblem Reading Glasses with Lanyard +1.75',
  'score': 0.12},
 {'value': 'Paper Mate Write Bros. Grip Mechanical Pencil, 0.7mm 5 ea (Pack of 2)',
  'score': 0.11},
 {'value': 'Christopher Radko Glass Plum Frosty Snowman Christmas Ornament #1017624',
  'score': 0.11},
 {'value': 'Is It To Late To Be Good Grinch Christmas Mens Tank Top Shirt',
  'score': 0.11}]
```

Figure 3-6. *The output*

Let's get recommendations using Manhattan similarity for CountVectorizer features.

```
#Manhattan Similarity
get_recommendation_cv(product_id, unique_df, similarity =
"manhattan", n=10)
```

Figure 3-7 shows the output of Manhattan similarity for CountVectorizer features.

```
[{'value': 'Stepping Stones', 'score': 43.0},
 {'value': 'Global Portuguese', 'score': 43.0},
 {'value': 'Polo Blue by Ralph Lauren', 'score': 43.0},
 {'value': 'Auburn Leathercrafters Tuscany Leather Dog Collar', 'score': 45.0},
 {'value': 'Leftover Salmon', 'score': 45.0},
 {'value': 'Good (Vinyl)', 'score': 45.0},
 {'value': 'Drunken Monkeys', 'score': 45.0},
 {'value': 'DuraTech Roof Support Trim', 'score': 47.0},
 {'value': 'Amerlite Niche Sealing Ring', 'score': 47.0},
 {'value': 'Learning and Performance in Corrections', 'score': 47.0}]
```

Figure 3-7. *The output*

Next, get recommendations using Euclidean similarity for CountVectorizer features.

```
#Euclidean Similarity
get_recommendation_cv(product_id, unique_df, similarity =
"euclidean", n=10)
```

Figure 3-8 shows the Euclidean output for CountVectorizer features.

```
[{'value': 'Polo Blue by Ralph Lauren', 'score': 9.0},
 {'value': 'Auburn Leathercrafters Tuscany Leather Dog Collar', 'score': 9.11},
 {'value': 'Global Portuguese', 'score': 9.11},
 {'value': 'Stepping Stones', 'score': 9.22},
 {'value': 'Always in My Heart', 'score': 9.22},
 {'value': 'Leftover Salmon', 'score': 9.22},
 {'value': 'Good (Vinyl)', 'score': 9.22},
 {'value': 'Drunken Monkeys', 'score': 9.22},
 {'value': 'Learning and Performance in Corrections', 'score': 9.43},
 {'value': 'Chasing Hamburg (Vinyl)', 'score': 9.43}]
```

Figure 3-8. *The output*

Build a Model Using TF-IDF Features

Using TF-IDF features, let's write a function that recommends the top ten most similar products.

```
# Comparing similarity to get the top matches using TF-IDF

def get_recommendation_tfidf(product_id, df, similarity, n=10):

    row = df.loc[df['Product Name'] == product_id]
    index = list(row.index)[0]
    description = row['desc_lowered'].loc[index]
```

```
    #Create vector using tfidf

    tfidf_matrix = tfidf_vec.fit_transform(desc_list)

    if similarity == "cosine":
        sim_matrix = cosine_similarity(tfidf_matrix)
        products = find_similarity(sim_matrix , index)

    elif similarity == "manhattan":
        sim_matrix = manhattan_distances(tfidf_matrix)
        products = find_manhattan_distance(sim_matrix , index)

    else:
        sim_matrix = euclidean_distances(tfidf_matrix)
        products = find_euclidean_distances(sim_matrix , index)

    return products
```

The input for this function is the same as what was used in the previous section. The recommendations are for the same product.

Next, get recommendations using cosine similarity for TF-IDF features.

```
# Cosine Similarity
get_recommendation_tfidf(product_id, unique_df, similarity =
"cosine", n=10)
```

Figure 3-9 shows the cosine output for TF-IDF features.

```
[{'value': 'Fancyleo Christmas Glasses Frames 2 Pack Glittered Eyeglasses Glasses Set No Lens Kids Family Xmas Party Ornamen
ts Gift',
  'score': 0.07},
 {'value': 'storefront christmas LED Decoration Light Gold Color Star Shape Vine Wedding Party event',
  'score': 0.05},
 {'value': '12ct Light Gunmetal Gray Shatterproof 4-Finish Christmas Ball Ornaments 4" (100mm)',
  'score': 0.05},
 {'value': '16 inches 40 inches "MERRY CHRISTMAS" Letter Foil Inflated Balloon Float Helium Aluminum Balloons for Christmas
Decoration',
  'score': 0.05},
 {'value': 'Is It To Late To Be Good Grinch Christmas Mens Tank Top Shirt',
  'score': 0.02},
 {'value': 'Christopher Radko Glass Plum Frosty Snowman Christmas Ornament #1017624',
  'score': 0.02},
 {'value': 'CMFUN Watercolor Brush Creative Flower Made with Ink Hand Painting for Your Designs Pillowcase 20x20 inch',
  'score': 0.02},
 {'value': 'SKIN DECAL FOR OtterBox Symmetry Samsung Galaxy S7 Case - Christmas Snowflake Blue Ornaments DECAL, NOT A CASE',
  'score': 0.02},
 {'value': "Santa's Workshop Illinois Mascot and Flag Nutcracker",
  'score': 0.02},
 {'value': 'The Holiday Aisle LED C7 Faceted Christmas Light Bulb',
  'score': 0.02}]
```

Figure 3-9. *The output*

To get recommendations using Manhattan similarity for TF-IDF features, change similarity to "manhattan".

```
#Manhattan Similarity
get_recommendation_tfidf(product_id, unique_df, similarity =
"manhattan", n=10)
```

To get recommendations using Euclidean similarity for TF-IDF features, change similarity to "euclidean".

```
#Euclidean Similarity
get_recommendation_tfidf(product_id, unique_df, similarity =
"euclidean", n=10)
```

Build a Model Using Word2vec Features

Using Word2vec features, let's write a function that recommends the top ten most similar products.

```
#  Comparing similarity to get the top matches using Word2vec
pretrained model

def get_recommendation_word2vec(product_id, df, similarity, n=10):

    row = df.loc[df['Product Name'] == product_id]
    input_index = list(row.index)[0]
    description = row['desc_lowered'].loc[input_index]

    #create vectors for each desc using word2vec
    vector_matrix = np.empty((len(desc_list), 300))
    for index, each_sentence in enumerate(desc_list):
        sentence_vector = np.zeros((300,))
        count  = 0
        for each_word in each_sentence.split():
            try:
                sentence_vector += word2vecModel[each_word]
                count += 1
            except:
                continue
```

```
        vector_matrix[index] = sentence_vector

    if similarity == "cosine":
        sim_matrix = cosine_similarity(vector_matrix)
        products = find_similarity(sim_matrix , input_index)

    elif similarity == "manhattan":
        sim_matrix = manhattan_distances(vector_matrix)
        products = find_manhattan_distance(sim_matrix , input_index)

    else:
        sim_matrix = euclidean_distances(vector_matrix)
        products = find_euclidean_distances(sim_matrix , input_index)

    return products
```

The input for this function is the same as what was used in the previous section. The recommendations are for the same product.

Let's get recommendations using Manhattan similarity for Word2vec features.

```
#Manhattan Similarity
get_recommendation_word2vec(product_id, unique_df, similarity =
"manhattan", n=10)
```

Figure 3-10 shows the Manhattan output for Word2vec features.

```
[{'value': 'storefront christmas LED Decoration Light Gold Color Star Shape Vine Wedding Party event',
  'score': 458.13},
 {'value': '8 1/2 x 14 Cardstock - Crystal Metallic (500 Qty.)',
  'score': 488.19},
 {'value': 'Cavalier Spaniel St. Patricks Day Shamrock Mouse Pad&#44; Hot Pad Or Trivet',
  'score': 497.0},
 {'value': "Call of the Wild Howling the Full Moon Women's Racerback Alpha Wolf",
  'score': 509.22},
 {'value': 'Fringe Table Skirt Purple 9 ft x 29 inches Pkg/1',
  'score': 516.08},
 {'value': 'Trend Enterprises T-83315 1.25 in. Holiday Pals & Peppermint Scratch N Sniff Stinky Stickers&#44; Large Round',
  'score': 522.0},
 {'value': "Allwitty 1039 - Women's T-Shirt Ipac Pistol Gun Apple Iphone Parody",
  'score': 525.03},
 {'value': 'Clear 18 Note Acrylic Box Musical Paperweight - Light My Fire',
  'score': 526.08},
 {'value': 'Handcrafted Ercolano Music Box Featuring "Luncheon of the Boating Party" by Renoir, Pierre Auguste - New YorkNew
York',
  'score': 527.88},
 {'value': 'Platinum 5 mm Comfort Fit Half Round Wedding Band - Size 9.5',
  'score': 528.08}]
```

Figure 3-10. *The output*

To get recommendations using cosine similarity for Word2vec features, change similarity to "cosine".

```
# Cosine Similarity
get_recommendation_word2vec(product_id, unique_df, similarity =
"cosine", n=10)
```

To get recommendations using Euclidean similarity for Word2vec features, change similarity to "euclidean".

```
#Euclidean Similarity
get_recommendation_word2vec(product_id, unique_df, similarity =
"euclidean", n=10)
```

Build a Model Using fastText Features

Using fastText features, let's write a function that recommends the top ten most similar products.

```
#  Comparing similarity to get the top matches using fastText
pretrained model

def get_recommendation_fasttext(product_id, df, similarity, n=10):

    row = df.loc[df['Product Name'] == product_id]
    input_index = list(row.index)[0]
    description = row['desc_lowered'].loc[input_index]

    #create vectors for each description using fasttext
    vector_matrix = np.empty((len(desc_list), 300))
    for index, each_sentence in enumerate(desc_list):
        sentence_vector = np.zeros((300,))
        count  = 0
        for each_word in each_sentence.split():
            try:
                sentence_vector += fasttext_model.wv[each_word]
                count += 1
            except:
                continue
```

```
        vector_matrix[index] = sentence_vector

    if similarity == "cosine":
        sim_matrix = cosine_similarity(vector_matrix)
        products = find_similarity(sim_matrix , input_index)

    elif similarity == "manhattan":
        sim_matrix = manhattan_distances(vector_matrix)
        products = find_manhattan_distance(sim_matrix , input_index)

    else:
        sim_matrix = euclidean_distances(vector_matrix)
        products = find_euclidean_distances(sim_matrix , input_index)

    return products
```

The input for this function is the same as what was used in the previous section. The recommendations are for the same product.

Let's get recommendations using cosine similarity for fastText features.

```
# Cosine Similarity
get_recommendation_fasttext(product_id, unique_df, similarity =
"cosine", n=10)
```

Figure 3-11 shows the cosine for fastText features output.

```
[{'value': 'All Weather Cornhole Bags - Set of 8', 'score': 0.95},
 {'value': 'American Foxhound Christmas Sticky Note Holder BB8433SN',
  'score': 0.95},
 {'value': '94" Bottom Width x 96 1/2" Top Width x 5 1/2"H x 1 3/4"P Stockton Crosshead',
  'score': 0.94},
 {'value': 'Business Essentials 8" x 8" x 5" Corrugated Mailers, 12-Pack',
  'score': 0.94},
 {'value': 'Efavormart Pack of 5 Premium 17" x 17" Washable Polyester Napkins Great for Wedding Party Restaurant Dinner Part
ies',
  'score': 0.94},
 {'value': '16 inches 40 inches "MERRY CHRISTMAS" Letter Foil Inflated Balloon Float Helium Aluminum Balloons for Christmas
Decoration',
  'score': 0.94},
 {'value': 'Ribbon Bazaar Double Faced Satin 2-1/4 inch Leaf Green 25 yards 100% Polyester Ribbon',
  'score': 0.94},
 {'value': 'Buckle-Down Pet Leash - Buffalo Plaid Black Green - 4 Feet Long - 1 2" Wide',
  'score': 0.94},
 {'value': "Diamond Clear Jewel Tone 3' Latex Balloon", 'score': 0.93},
 {'value': '48" Rect Resin Table & 6x14" Chairs - Sand', 'score': 0.93}]
```

Figure 3-11. *The output*

To get recommendations using Manhattan similarity for fastText features, change similarity to "manhattan".

```
#Manhattan Similarity
get_recommendation_fasttext(product_id, unique_df, similarity =
"manhattan", n=10)
```

To get recommendations using Euclidean similarity for fastText features, change similarity to "euclidean".

```
#Euclidean Similarity
get_recommendation_fasttext(product_id, unique_df, similarity =
"euclidean", n=10)
```

Build a Model Using GloVe Features

Using GloVe features, let's write a function that recommends the top ten most similar products.

```
#  Comparing similarity to get the top matches using GloVe pretrained model

def get_recommendation_glove(product_id, df, similarity, n=10):

    row = df.loc[df['Product Name'] == product_id]
    input_index = list(row.index)[0]
    description = row['desc_lowered'].loc[input_index]

    #using glove embeddings to create vectors
    vector_matrix = np.empty((len(desc_list), 300))
    for index, each_sentence in enumerate(desc_list):
        sentence_vector = np.zeros((300,))
        count  = 0
        for each_word in each_sentence.split():
            try:
                sentence_vector += glove_model[each_word]
                count += 1
```

```
        except:
            continue

    vector_matrix[index] = sentence_vector

  if similarity == "cosine":
     sim_matrix = cosine_similarity(vector_matrix)
     products = find_similarity(sim_matrix , input_index)

  elif similarity == "manhattan":
     sim_matrix = manhattan_distances(vector_matrix)
     products = find_manhattan_distance(sim_matrix , input_index)

  else:
     sim_matrix = euclidean_distances(vector_matrix)
     products = find_euclidean_distances(sim_matrix , input_index)

  return products
```

The input for this function is the same as what was used in the previous section. The recommendations are for the same product.

Next, get recommendations using Euclidean similarity for GloVe features.

```
#Euclidean Similarity
get_recommendation_glove(product_id, unique_df, similarity =
"euclidean", n=10)
```

Figure 3-12 shows the Euclidean output for GloVe features.

```
[{'value': 'Spiral Birthday Candles, 36 Count', 'score': 19.13},
 {'value': 'Just Artifacts Gold Glitter Letter B', 'score': 19.92},
 {'value': 'Giant 36in. Purple Balloons (Set of 2)', 'score': 21.02},
 {'value': '(2-Pack) StealthShields Tablet Screen Protector for Lenovo IdeaPad Yoga 11 (U...',
  'score': 22.99},
 {'value': 'Ganma Superheroes Ordinary Life Case For Samsung Galaxy Note 5 Hard Case Cover',
  'score': 23.2},
 {'value': 'Platinum 5 mm Comfort Fit Half Round Wedding Band - Size 9.5',
  'score': 23.21},
 {'value': 'IN-70/65 Blue Paper Streamers 2PK', 'score': 23.46},
 {'value': "New Way 075 - Men's Sleeveless Fbi Female Body Inspector",
  'score': 23.91},
 {'value': 'Coral Parchment Treat Bags', 'score': 24.03},
 {'value': '031 - Unisex Long-Sleeve T-Shirt Disobey V For Vendetta Anonymous Fawkes Mask',
  'score': 24.23}]
```

Figure 3-12. *The output*

To get recommendations using cosine similarity for GloVe features, change similarity to "cosine".

```
# Cosine Similarity
get_recommendation_glove(product_id, unique_df, similarity =
"cosine", n=10)
```

To get recommendations using Manhattan similarity for GloVe features, change similarity to "manhattan".

```
#Manhattan Similarity
get_recommendation_glove(product_id, unique_df, similarity =
"manhattan", n=10)
```

The purpose of a co-occurrence matrix is to present the number of times each word appears in the same context.

"Roses are red. The sky is blue." Figure 3-13 shows these words in a co-occurrence matrix.

	Roses	are	red	Sky	is	blue
Roses	1	1	1	0	0	0
are	1	1	1	0	0	0
red	1	1	1	0	0	0
Sky	0	0	0	1	1	1
is	0	0	0	1	1	1
Blue	0	0	0	1	1	1

Figure 3-13. *Co-occurrence matrix*

The disadvantage of this method is it is time-consuming; in real-time, it is rarely used. Since it is time-consuming, let's take a few records from the dataset and implement them.

```
# create cooccurence matrix

#preprocessing
df = df.head(250)
# Combining Product and Description
df['Description'] = df['Product Name'] + ' ' +df['Description']
unique_df = df.drop_duplicates(subset=['Description'], keep='first')
```

```
unique_df['desc_lowered'] = unique_df['Description'].apply(lambda x:
x.lower())
unique_df['desc_lowered'] = unique_df['desc_lowered'].apply(lambda x:
re.sub(r'[^\w\s]', '', x))
desc_list = list(unique_df['desc_lowered'])

co_ocr_vocab = []
for i in desc_list:
    [co_ocr_vocab.append(x) for x in i.split()]

co_occur_vector_matrix = np.zeros((len(co_ocr_vocab), len(co_ocr_vocab)))

for _, sent in enumerate(desc_list):
    words = sent.split()
    for index, word in enumerate(words):
        if index != len(words)-1:
            co_occur_vector_matrix[co_ocr_vocab.index(word)][co_ocr_vocab.
            index(words[index+1])] += 1
```

Build a Model Using a Co-occurrence Matrix

Using co-occurrence features, let's write a function that recommends the top ten most similar products.

```
#  Comparing similarity to get the top matches using cooccurence matrix

def get_recommendation_coccur(product_id, df, similarity, n=10):

    row = df.loc[df['Product Name'] == product_id]
    input_index = list(row.index)[0]
    description = row['desc_lowered'].loc[input_index]

    vector_matrix = np.empty((len(desc_list), len(co_ocr_vocab)))
    for index, each_sentence in enumerate(desc_list):
        sentence_vector = np.zeros((len(co_ocr_vocab),))
        count  = 0
        for each_word in each_sentence.split():
            try:
```

```
                sentence_vector += co_occur_vector_matrix[co_ocr_vocab.
                index(each_word)]
                count += 1

        except:
            continue

    vector_matrix[index] = sentence_vector/count

if similarity == "cosine":
    sim_matrix = cosine_similarity(vector_matrix)
    products = find_similarity(sim_matrix , index)

elif similarity == "manhattan":
    sim_matrix = manhattan_distances(vector_matrix)
    products = find_manhattan_distance(sim_matrix , index)

else:
    sim_matrix = euclidean_distances(vector_matrix)
    products = find_euclidean_distances(sim_matrix , index)

return products
```

Next, get recommendations using Euclidean similarity for co-occurrence features.

```
#Euclidean Similarity
get_recommendation_coccur(product_id, unique_df, similarity =
"euclidean", n=10)
```

Figure 3-14 shows the Euclidean output for the co-occurrence matrix.

```
[{'value': 'Toddler Kid Boys Girls Lightweight Breathable Trendy Slip-on Sneaker (6M US Toddler, Red)',
  'score': 2.17},
 {'value': "Pull-Ups Girls' Learning Designs Training Pants (Choose Pant Size and Count)",
  'score': 2.32},
 {'value': 'Medi Comfort Closed Toe Knee Highs -15-20 mmHg Reg',
  'score': 2.35},
 {'value': "JustVH Women's Solid Henley V-Neck Casual Blouse Pleated Button Tunic Shirt Top",
  'score': 2.53},
 {'value': "Dr. Comfort Paradise Women's Casual Shoe: 4.5 X-Wide (E-2E) Black Velcro",
  'score': 2.73},
 {'value': 'Box Packaging White Deluxe Literature Mailer, 50/Bundle',
  'score': 2.8},
 {'value': 'Ebe Reading Glasses Mens Womens Amber Red Oval Round Full Frame Anti Glare grade ckbdp9118',
  'score': 2.81},
 {'value': 'Nail DIP Powder, Classic Color Collection, Dipping Acrylic For Any Kit or System by DipWell (CL - 58)',
  'score': 2.85},
 {'value': "Women's Breeze Walker", 'score': 2.94},
 {'value': 'Bare Nature Vitamin Iced Tea - Guava Pineapple, 20 Fl. Oz. Bottles, 12 Ct',
  'score': 3.03}]
```

Figure 3-14. *The output*

To get recommendations using cosine similarity, change similarity to "cosine".

```
# Cosine Similarity
get_recommendation_coccur(product_id, unique_df, similarity =
"cosine", n=10)
```

To get recommendations using Manhattan similarity, change similarity to "manhattan".

```
#Manhattan Similarity
get_recommendation_coccur(product_id, unique_df, similarity =
"manhattan", n=10)
```

Summary

In this chapter, you learned how to build a content-based model using text data from the data preparation to the recommendations to the users. You saw models built using several NLP techniques. Using word embeddings is a much better option, given they have the power to capture context and semantics.

CHAPTER 4

Collaborative Filtering

Collaborative filtering is a very popular method in recommendation engines. It is the predictive process behind the suggestions provided by these systems. It processes and analyzes customers' information and suggests items they will likely appreciate.

Collaborative filtering algorithms use a customer's purchase history and ratings to find similar customers and then suggest items that they liked.

Figure 4-1 explains collaborative filtering at a high level.

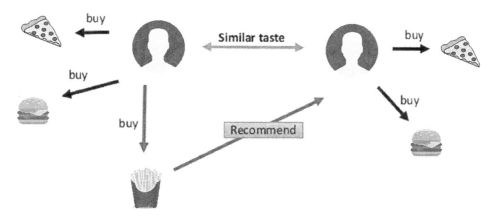

Figure 4-1. *Collaborative filtering explained*

For example, to find a new movie or show to watch, you can ask your friends for suggestions since you all share similar tastes in content. The same concept is used in collaborative filtering, where user-user similarity finds similar users to get recommendations based on each other's likes.

There are two types of collaborative filtering methods—user-to-user and item-to-item. They are explored in the upcoming sections. This chapter looks at the implementation of these two methods using cosine similarity before diving into implementing the more popularly used KNN-based algorithm for collaborative filtering.

© Akshay Kulkarni, Adarsha Shivananda, Anoosh Kulkarni, V Adithya Krishnan 2023
A. Kulkarni et al., *Applied Recommender Systems with Python*, https://doi.org/10.1007/978-1-4842-8954-9_4

Implementation

The following installs the surprise library.

```
!pip install scikit-surprise
```

The following imports basic libraries.

```
import pandas as pd
import numpy as np
import seaborn as sns
import matplotlib.pyplot as plt
%matplotlib inline
import random
from IPython.display import Image
```

The following imports the KNN algorithm and csr_matrix for KNN data preparation.

```
from scipy.sparse import csr_matrix
from sklearn.neighbors import NearestNeighbors
```

The following calculates cosine similarity by importing cosine_similarity.

```
from sklearn.metrics.pairwise import cosine_similarity
```

Let's import surprise.Reader and surprise.Dataset for surprise data preparation.

```
from surprise import Reader, Dataset
```

Next, import surprise.model_selection functions for surprise model customizations.

```
from surprise.model_selection import train_test_split, cross_validate,
GridSearchCV
```

Then, import algorithms from the surprise package.

```
from surprise.prediction_algorithms import CoClustering
from surprise.prediction_algorithms import NMF
```

Finally, import accuracy to get metrics such as root-mean-square error (RMSE) and *mean absolute error* (MAE).

```
from surprise import accuracy
```

Data Collection

This chapter uses a custom dataset that has been masked. Download the dataset from the GitHub link.

The following reads the data.

```
data = pd.read_excel('Rec_sys_data.xlsx',encoding= 'unicode_escape')
data.head()
```

Figure 4-2 shows the DataFrame.

	InvoiceNo	StockCode	Quantity	InvoiceDate	DeliveryDate	Discount%	ShipMode	ShippingCost	CustomerID
0	536365	84029E	6	2010-12-01 08:26:00	2010-12-02 08:26:00	0.57	ExpressAir	30.12	17850
1	536365	71053	6	2010-12-01 08:26:00	2010-12-03 08:26:00	0.15	Regular Air	15.22	17850
2	536365	21730	6	2010-12-01 08:26:00	2010-12-03 08:26:00	0.57	Regular Air	15.22	17850
3	536365	84406B	8	2010-12-01 08:26:00	2010-12-02 08:26:00	0.15	ExpressAir	30.12	17850
4	536365	22752	2	2010-12-01 08:26:00	2010-12-02 08:26:00	0.47	ExpressAir	30.12	17850

Figure 4-2. *Input data*

About the Dataset

The following is the data dictionary for the dataset; it has nine features (columns).

- InvoiceNo: The invoice number of a particular transaction
- StockCode: The unique identifier for a particular item
- Quantity: The quantity of that item bought by the customer
- InvoiceDate: The date and time when the transaction was made
- DeliveryDate: The date and time when the delivery happened
- Discount%: Percentage of discount on the purchased item
- ShipMode: Mode of shipping
- ShippingCost: Cost of shipping that item
- CustomerID: The unique identifier of a particular customer

The following checks the size of the data.

```
data.shape
```

```
(272404, 9)
```

The dataset has a total of 272,404 unique transactions in its nine columns.

Let's check if there are any null values because a clean dataset is required for further analysis.

```
data.isnull().sum().sort_values(ascending=False)
```

The following is the output.

```
CustomerID      0
ShippingCost    0
ShipMode        0
Discount%       0
DeliveryDate    0
InvoiceDate     0
Quantity        0
StockCode       0
InvoiceNo       0
dtype: int64
```

The data is clean with no nulls in any columns. Further preprocessing is not required in this case.

If there were any NaNs or nulls in the data, they were dropped using the following.

```
data1 = data.dropna()
```

Now let's check for any data abnormalities by describing the data.

```
data1.describe()
```

Figure 4-3 describes data1.

	InvoiceNo	Quantity	Discount%	ShippingCost	CustomerID
count	272404.000000	272404.000000	272404.000000	272404.000000	272404.000000
mean	553740.733319	13.579536	0.300164	17.032280	15284.323523
std	9778.082879	149.136756	0.176173	10.011102	1714.478624
min	536365.000000	1.000000	0.000000	5.810000	12346.000000
25%	545312.000000	2.000000	0.150000	5.810000	13893.000000
50%	553902.000000	6.000000	0.300000	15.220000	15157.000000
75%	562457.000000	12.000000	0.450000	30.120000	16788.000000
max	569629.000000	74215.000000	0.600000	30.120000	18287.000000

Figure 4-3. *data1*

There aren't any negative values in the Quantity column, but if there were, those records would need to be dropped since it's a data abnormality.

Let's change the StockCode column datatype to string to maintain the same type across all rows.

```
data1.StockCode = data1.StockCode.astype(str)
```

Memory-Based Approach

Let's examine the most basic approach to implementing collaborative filtering: the memory-based approach. This approach uses simple arithmetic operations or metrics to calculate the similarities between two users or two items to group them. For example, to find user-user relations, both users' historically liked items are used to find the similarity metric, that measures how similar the two users are.

Cosine similarity is a common similarity metric. Euclidean distance and Pearson's correlation are other popular metrics. A metric is considered geometric if the row (column) of a given user (item) is treated as a vector or a matrix. In cosine similarity, the similarity of two users (say) is measured as the cosine of the angle between the vectors of the two users. For users A and B, the cosine similarity is given by the formula shown in Figure 4-4.

$$\text{similarity} = \cos(\theta) = \frac{\mathbf{A} \cdot \mathbf{B}}{\|\mathbf{A}\|\|\mathbf{B}\|}$$

Figure 4-4. *Cosine similarity formula*

This approach is easy to implement and understand because no model training or heavy optimization algorithms are involved. However, its performance degrades when there is sparse data. For this method to work precisely, huge amounts of clean data on multiple users and items are required, which hinders the scalability of this approach for most real-world applications.

The memory-based approach is further divided into user-to-user-based and item-to-item-based collaborative filtering.

The implementation of both methods is explored in this chapter.

Figure 4-5 illustrates user-based and item-based filtering.

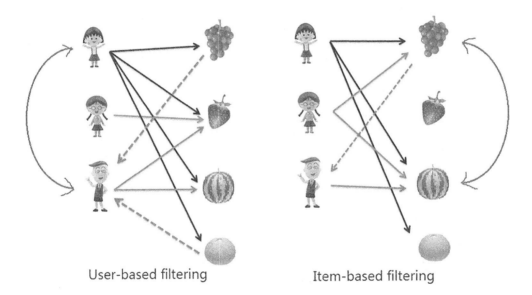

User-based filtering   Item-based filtering

Figure 4-5. *User-based and item-based collaborative filtering*

User-to-User Collaborative Filtering

User-to-user-based collaborative filtering recommends items that a particular user might like by finding similar users, using purchase history or ratings on various items, and then suggesting the items liked by these similar users.

Here, a matrix is formed to describe the behavior of all the users (purchase history in our example) corresponding to all the items. Using this matrix, you can calculate the similarity metrics (cosine similarity) between users to formulate user-user relations. These relations help find users similar to a given user and recommend the items bought by these similar users.

Implementation

Let's first create a data matrix covering purchase history. It contains all customer IDs for all items (whether a customer has purchased an item or not).

```
purchase_df = (data1.groupby(['CustomerID', 'StockCode'])['Quantity'].
sum().unstack().reset_index().fillna(0).set_index('CustomerID'))
purchase_df.head()
```

Figure 4-6 shows the purchase data matrix.

StockCode	10002	10080	10120	10125	10133	10135	11001	15030	15034	15036	...	90214R	90214S	90214V	90214Y	BANK CHARGES	C2	DOT	M	PADS	F
CustomerID																					
12346	0.0	0.0	0.0	0.0	0.0	0.0	0.0	0.0	0.0	0.0	...	0.0	0.0	0.0	0.0	0.0	0.0	0.0	0.0	0.0
12347	0.0	0.0	0.0	0.0	0.0	0.0	0.0	0.0	0.0	0.0	...	0.0	0.0	0.0	0.0	0.0	0.0	0.0	0.0	0.0
12348	0.0	0.0	0.0	0.0	0.0	0.0	0.0	0.0	0.0	0.0	...	0.0	0.0	0.0	0.0	0.0	0.0	0.0	0.0	0.0
12350	0.0	0.0	0.0	0.0	0.0	0.0	0.0	0.0	0.0	0.0	...	0.0	0.0	0.0	0.0	0.0	0.0	0.0	0.0	0.0
12352	0.0	0.0	0.0	0.0	0.0	0.0	0.0	0.0	0.0	0.0	...	0.0	0.0	0.0	0.0	0.0	0.0	0.0	3.0	0.0

5 rows × 3538 columns

Figure 4-6. *Purchase data matrix*

The data matrix shown in Figure 4-6 reveals the total quantity purchased by each user against each item. Only information about whether the item was bought or not by the user is needed, not the quantity.

Thus, an encoding of 0 or 1 is used, where 0 is not purchased, and 1 is purchased.

Let's first write a function for encoding the data matrix.

```
def encode_units(x):
    if x < 1: # If the quantity is less than 1
        return 0 # Not purchased
    if x >= 1: # If the quantity is greater than 1
        return 1 # Purchased
```

Next, apply this function to the data matrix.

```
purchase_df = purchase_df.applymap(encode_units)
purchase_df.head()
```

Figure 4-7 shows the purchase data matrix after encoding.

StockCode	10002	10080	10120	10125	10133	10135	11001	15030	15034	15036	...	90214R	90214S	90214V	90214Y	BANK CHARGES	C2	DOT	M	PADS	P(
CustomerID																					
12346	0	0	0	0	0	0	0	0	0	0	...	0	0	0	0	0	0	0	0	0	
12347	0	0	0	0	0	0	0	0	0	0	...	0	0	0	0	0	0	0	0	0	
12348	0	0	0	0	0	0	0	0	0	0	...	0	0	0	0	0	0	0	0	0	
12350	0	0	0	0	0	0	0	0	0	0	...	0	0	0	0	0	0	0	0	0	
12352	0	0	0	0	0	0	0	0	0	0	...	0	0	0	0	0	0	0	1	0	

5 rows × 3538 columns

Figure 4-7. *Purchase data matrix after encoding*

The purchase data matrix reveals the behavior of customers across all items. This matrix finds the user similarity scores matrix, and the similarity metric uses cosine similarity. The user similarity score matrix has user-to-user similarity for each user pair.

First, let's apply cosine_similarity to the purchase data matrix.

```
user_similarities = cosine_similarity(purchase_df)
```

Now, let's store the user similarity scores in a DataFrame (i.e., the similarity scores matrix).

```
user_similarity_data = pd.DataFrame(user_similarities,index=purchase_
df.index,columns=purchase_df.index)
user_similarity_data.head()
```

Figure 4-8 shows the user similarity scores data matrix.

CustomerID	12346	12347	12348	12350	12352	12353	12354	12355	12356	12358	...	18269	18270	18272	18273	18278
CustomerID																
12346	1.0	0.000000	0.000000	0.000000	0.000000	0.0	0.000000	0.000000	0.000000	0.000000	...	0.000000	0.000000	0.114708	0.0	0.000000
12347	0.0	1.000000	0.070632	0.053567	0.048324	0.0	0.029001	0.091885	0.075845	0.000000	...	0.041739	0.000000	0.050869	0.0	0.036811
12348	0.0	0.070632	1.000000	0.051709	0.031099	0.0	0.027995	0.118262	0.146427	0.061546	...	0.000000	0.000000	0.024456	0.0	0.000000
12350	0.0	0.053567	0.051709	1.000000	0.035377	0.0	0.000000	0.000000	0.033315	0.070014	...	0.000000	0.000000	0.027821	0.0	0.000000
12352	0.0	0.048324	0.031099	0.035377	1.000000	0.0	0.095765	0.040456	0.100180	0.084215	...	0.110264	0.065233	0.133855	0.0	0.000000

5 rows × 3647 columns

Figure 4-8. *User similarity scores DataFrame*

The similarity score values are between 0 to 1, where values closer to 0 represent less similar, and values closer to 1 represent more similar customers.

Using this user similarity scores data, let's get recommendations for a given user. Create a function for this.

```
def fetch_similar_users(user_id,k=5):
    # separating data rows for the entered user id
    user_similarity = user_similarity_data[user_similarity_data.index ==
    user_id]

    # a data of all other users
    other_users_similarities = user_similarity_data[user_similarity_data.
    index != user_id]

    # calcuate cosine similarity between user and each other user
    similarities = cosine_similarity(user_similarity,other_users_
    similarities)[0].tolist()

    user_indices = other_users_similarities.index.tolist()

    index_similarity_pair = dict(zip(user_indices, similarities))

    # sort by similarity
    sorted_index_similarity_pair = sorted(index_similarity_pair.
    items(),reverse=True)

    top_k_users_similarities = sorted_index_similarity_pair[:k]
    similar_users = [u[0] for u in top_k_users_similarities]

    print('The users with behaviour similar to that of user {0} are:'.
    format(user_id))
    return similar_users
```

This function separates the selected user from all other users and then takes a cosine similarity of the selected user with all users to find similar users. Return the top k similar users (by CustomerID) to our selected user.

For example, let's find the similar to user 12347.

```
similar_users = fetch_similar_users(12347)
```

similar_users

The following is the output.

The users with behavior similar to that of user 12347 are:
[18287, 18283, 18282, 18281, 18280]

As expected, the default five users are similar to user 12347.

Now, let's get the recommendations by showing the items bought by similar users.

Write another function to get similar user recommendations.

```
def simular_users_recommendation(userid):

    similar_users = fetch_similar_users(userid)

    #obtaining all the items bought by similar users
    simular_users_recommendation_list = []
    for j in similar_users:
        item_list = data1[data1["CustomerID"]==j]['StockCode'].to_list()
        simular_users_recommendation_list.append(item_list)

    #this gives us multi-dimensional list
    # we need to flatten it
    flat_list = []
    for sublist in simular_users_recommendation_list:
        for item in sublist:
            flat_list.append(item)
    final_recommendations_list = list(dict.fromkeys(flat_list))

    # storing 10 random recommendations in a list
    ten_random_recommendations = random.sample(final_recommendations_
    list, 10)

    print('Items bought by Similar users based on Cosine Similarity')

    #returning 10 random recommendations
    return ten_random_recommendations
```

This function gets the similar users for the given customer (ID) and obtains a list of all the items bought by these similar users. This list is then flattened to get a final list of unique items, from which shows randomly chosen ten recommended items for a given user.

Using this function on user 12347 to get recommendations results in the following suggestions.

```
simular_users_recommendation(12347)
```

The following is the output.

```
The users with behavior similar to that of user 12347 are:
Items bought by Similar users based on Cosine Similarity
['21967', '21908', '21154', '20723', '23296', '22271', '22746', '22355',
'22554', '23199']
```

User 12347 had ten suggestions from the items bought by similar users.

Item-to-Item Collaborative Filtering

Item-to-item based collaborative filtering recommends items that a particular user might like by finding items similar to ones they already purchased and then creating a matrix profile for each item. Purchase history or user ratings are also used.

A matrix is formed to describe the behavior of all the users (purchase history in our example) corresponding to all the items. This matrix helps calculate the similarity metrics (cosine similarity) between items to formulate the item-item relations. This relation is used to recommend items similar to those previously purchased by the selected user.

Implementation

Following the initial steps used in user-to-user collaborative filtering methods, let's first create the data matrix, which contains all the item IDs across their purchase history (i.e., quantity purchased by each customer).

```
items_purchase_df = (data1.groupby(['StockCode','CustomerID'])['Quantity'].
sum().unstack().reset_index().fillna(0).set_index('StockCode'))
items_purchase_df.head()
```

The following is the output.

Figure 4-9 shows the item purchase data matrix.

CustomerID	12346	12347	12348	12350	12352	12353	12354	12355	12356	12358	...	18269	18270	18272	18273	18278	18280	18281	18282	18283
StockCode																				
10002	0.0	0.0	0.0	0.0	0.0	0.0	0.0	0.0	0.0	0.0	...	0.0	0.0	0.0	0.0	0.0	0.0	0.0	0.0	0.0
10080	0.0	0.0	0.0	0.0	0.0	0.0	0.0	0.0	0.0	0.0	...	0.0	0.0	0.0	0.0	0.0	0.0	0.0	0.0	0.0
10120	0.0	0.0	0.0	0.0	0.0	0.0	0.0	0.0	0.0	0.0	...	0.0	0.0	0.0	0.0	0.0	0.0	0.0	0.0	0.0
10125	0.0	0.0	0.0	0.0	0.0	0.0	0.0	0.0	0.0	0.0	...	0.0	0.0	0.0	0.0	0.0	0.0	0.0	0.0	0.0
10133	0.0	0.0	0.0	0.0	0.0	0.0	0.0	0.0	0.0	0.0	...	0.0	0.0	0.0	0.0	0.0	0.0	0.0	0.0	0.0

5 rows × 3647 columns

Figure 4-9. *Item purchase data matrix*

This data matrix shows the total quantity purchased by each user against each item. But the only information needed is whether the user bought the item.

Thus, an encoding of 0 or 1 is used, where 0 is not purchased, and 1 is purchased. Use the same encode_units function created earlier.

```
items_purchase_df = items_purchase_df.applymap(encode_units)
```

The items purchase data matrix reveals the behavior of customers across all items. Let's use this matrix to find item similarity scores with the cosine similarity metric. The item similarity score matrix has item-to-item similarity for each item pair.

First, let's apply cosine_similarity to the item purchase data matrix.

```
item_similarities = cosine_similarity(items_purchase_df)
```

Now, let's store the item similarity scores in a DataFrame (i.e., the similarity scores matrix).

```
item_similarity_data = pd.DataFrame(item_similarities,index=items_purchase_
df.index,columns=items_purchase_df.index)
item_similarity_data.head()
```

Figure 4-10 shows the item similarity scores data matrix.

StockCode	10002	10080	10120	10125	10133	10135	11001	15030	15034	15036	...	90214R	90214S	90214V	90214Y	BAN CHARGE
StockCode																
10002	1.000000	0.000000	0.108821	0.094281	0.062932	0.091902	0.110096	0.059761	0.083771	0.096449	...	0.0	0.0	0.0	0.0	0.
10080	0.000000	1.000000	0.000000	0.043033	0.028724	0.067116	0.000000	0.000000	0.076472	0.044023	...	0.0	0.0	0.0	0.0	0.
10120	0.108821	0.000000	1.000000	0.068399	0.068483	0.026669	0.079872	0.086711	0.121547	0.034986	...	0.0	0.0	0.0	0.0	0.
10125	0.094281	0.043033	0.068399	1.000000	0.044499	0.051988	0.051900	0.000000	0.039490	0.034100	...	0.0	0.0	0.0	0.0	0.
10133	0.062932	0.028724	0.068483	0.044499	1.000000	0.266043	0.051964	0.075218	0.079078	0.053110	...	0.0	0.0	0.0	0.0	0.

5 rows × 3538 columns

Figure 4-10. *Item similarity scores DataFrame*

The similarity score values are between 0 and 1, where values closer to 0 represent less similarity and values closer to 1 represent more similar items.

Using this item similarity score data, let's get recommendations for a given user. The following creates a function for this.

```
def fetch_similar_items(item_id,k=10):
    # separating data rows of the selected item
    item_similarity = item_similarity_data[item_similarity_data.index ==
    item_id]

    # a data of all other items
    other_items_similarities = item_similarity_data[item_similarity_data.
    index != item_id]

    # calculate cosine similarity between selected item with other items
    similarities = cosine_similarity(item_similarity,other_items_
    similarities)[0].tolist()

    # create list of indices of these items
    item_indices = other_items_similarities.index.tolist()

    # create key/values pairs of item index and their similarity
    index_similarity_pair = dict(zip(item_indices, similarities))

    # sort by similarity
    sorted_index_similarity_pair = sorted(index_similarity_pair.items())

    # grab k items from the top
    top_k_item_similarities = sorted_index_similarity_pair[:k]
    similar_items = [u[0] for u in top_k_item_similarities]
```

```
print('Similar items based on purchase behaviour (item-to-item
collaborative filtering)')
return similar_items
```

This function separates the selected item from all other items and then takes a cosine similarity of the selected item with all items to find the similarities. Return the top k similar items (StockCodes) to our selected item.

For example, let's find similar items for user 12347.

```
similar_items = fetch_similar_items('10002')
similar_items
```

The following is the output.

```
Similar items based on purchase behavior (item-to-item collaborative
filtering)
['10080',
 '10120',
 '10123C',
 '10124A',
 '10124G',
 '10125',
 '10133',
 '10135',
 '11001',
 '15030']
```

As expected, you see the default ten similar items to item 10002.

Now let's get the recommendations by showing similar items to those bought by a particular user.

Write another function to get similar item recommendations.

```
def simular_item_recommendation(userid):

    simular_items_recommendation_list = []

    #obtaining all the similar items to items bought by user
    item_list = data1[data1["CustomerID"]==userid]['StockCode'].to_list()
```

```
for item in item_list:
    similar_items = fetch_similar_items(item)
    simular_items_recommendation_list.append(item_list)

#this gives us multi-dimensional list
# we need to flatten it
flat_list = []
for sublist in simular_items_recommendation_list:
    for item in sublist:
        flat_list.append(item)
final_recommendations_list = list(dict.fromkeys(flat_list))

# storing 10 random recommendations in a list
ten_random_recommendations = random.sample(final_recommendations_
list, 10)

print('Similar Items bought by our users based on Cosine Similarity')

#returning 10 random recommendations
return ten_random_recommendations
```

This function gets the list of similar items for all previously bought items by our given customer (ID). This list is then flattened to get a final list of unique items, from which randomly chosen ten items as recommendations for our given user are shown.

Again, trying this function on user 12347 to get the recommendations for that user results in the following suggestions.

```
simular_item_recommendation(12347)
```

The following is the output.

```
Similar Items bought by our users based on Cosine Similarity
['22196',
 '22775',
 '22492',
 '23146',
 '22774',
 '21035',
 '16008',
```

```
'21041',
'23316',
'22550']
```

User 12347 has ten suggestions that are similar to items previously bought.

KNN-based Approach

You have learned the basics of collaborative filtering and implementing user-to-user and item-to-item filtering. Now let's dive into machine learning-based approaches, which are more robust and popular in building recommendation systems.

Machine Learning

Machine learning is a machine's capability to learn from experience (data) and make meaningful predictions without being explicitly programmed. It is a subfield of artificial intelligence that deals with building systems that can learn from data. The objective is to make computers learn on their own without any intervention from humans.

There are three primary machine learning categories.

Supervised Learning

In supervised learning, labeled training data is leveraged to derive the pattern or function and make a model or machine learn. Data consists of a dependent variable (Target label) and the independent variables or predictors. The machine tries to learn the function of labeled data and predict the output of unseen data.

Unsupervised Learning

In unsupervised learning, a machine learns the hidden pattern without leveraging labeled data, so training doesn't happen. These algorithms learn to capture patterns based on similarities or distances between data points.

Reinforcement Learning

Reinforcement learning is the process of maximizing a reward by taking action. The algorithms learn how to reach a goal through experience.

Figure 4-11 explains all the categories and subcategories.

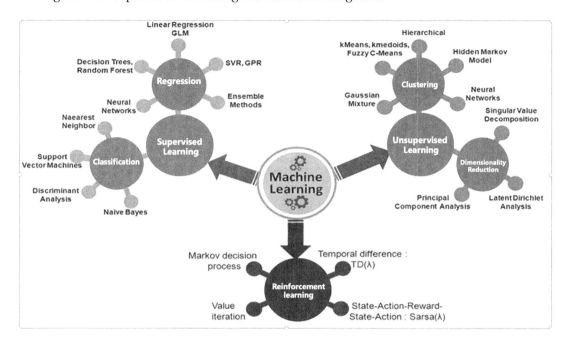

Figure 4-11. *Machine learning categories*

Supervised Learning

There are two types of supervised learning: regression and classification.

Regression

Regression is a statistical predictive modeling technique that finds the relationship between a dependent variable and one or more independent variables. Regression is used when the dependent variable is continuous; prediction can take any numerical value.

Popular regression algorithms include linear regression, decision tree, random forest, SVM, LightGBM, and XGBoost.

Classification

Classification is a supervised machine learning technique in which the dependent or output variable is categorical; for example, spam/ham, churn/not churned, and so on.

- In binary classification, it's either yes or no. There is no third option; for example, the customer can churn or not churn from a given business.

- In multiclass classification, the labeled variable can be multiclass, for example, product categorization of an e-commerce website.

Logistic regression, k-nearest neighbor, decision tree, random forest, SVM, LightGBM, and XGBoost are popular classification algorithms.

K-Nearest Neighbor

The k-nearest neighbor (KNN) algorithm is a supervised machine learning model that is used for both classification and regression problems. It is a very robust algorithm that is easy to implement and interpret and uses less calculation time. Labeled data is needed since it's a supervised learning algorithm.

Figure 4-12 explains KNN algorithms.

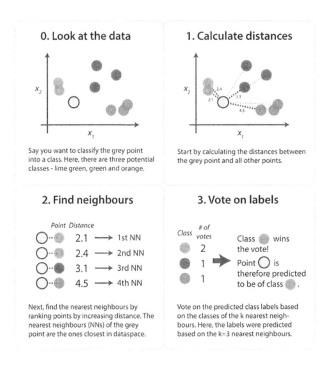

Figure 4-12. *KNN algorithm explained*

Now let's try implementing a simple KNN model on purchase_df, created in user-to-user filtering. This approach follows similar steps you have seen before (i.e., base recommendations from a list of items purchased by similar users). The difference is that a KNN model finds similar users (for a given user).

Implementation

Before passing our sparse matrix (i.e., purchase_df) into KNN, it must be converted into a CSR matrix.

CSR divides a sparse matrix into three separate arrays.

- values

- extent of rows

- index of columns

So, let's convert the sparse matrix into a CSR matrix.

```
purchase_matrix = csr_matrix(purchase_df.values)
```

Next, create the KNN model using the Euclidean distance metric.

```
knn_model = NearestNeighbors(metric = 'euclidean', algorithm = 'brute')
```

Once the model is created, fit it on the data/matrix.

```
knn_model.fit(purchase_matrix)
```

Figure 4-13 shows the fitted KNN model.

```
▼               NearestNeighbors
NearestNeighbors(algorithm='brute', metric='euclidean')
```

Figure 4-13. *Fitted KNN model*

Now that the KNN model is in place, let's write a function to fetch similar users using the model.

```
def fetch_similar_users_knn(purchase_df,query_index):

    # Creating empty list where we will store user id of similar users
```

```
simular_users_knn = []

# Storing the distance and index of nearest neighbor
distances, indices = knn_model.kneighbors(purchase_df.iloc[query_
index,:].values.reshape(1, -1), n_neighbors = 5)
for i in range(0, len(distances.flatten())):
    if i == 0:
        print('Recommendations for {0}:\n'.format(purchase_
        df.index[query_index]))
    else:
        print('{0}: {1}, with distance of {2}:'.format(i, purchase_
        df.index[indices.flatten()[i]], distances.flatten()[i]))

        simular_users_knn.append( purchase_df.index[indices.
        flatten()[i]])
```

This function first calculates the distances and indices of the five nearest neighbors using our KNN model's function. This output is then processed, and a list of similar users alone is returned. Instead of user_id as the input, take the index in the DataFrame.

Let's test this out for index 1497.

```
fetch_similar_users_knn(purchase_df,1497)
```

The following is the output.

```
Recommendations for 14729:

1: 16917, with distance of 8.12403840463596:
2: 16989, with distance of 8.12403840463596:
3: 15124, with distance of 8.12403840463596:
4: 12897, with distance of 8.246211251235321:
```

```
simular_users_knn
```

The following is the output.

```
[16917, 16989, 15124, 12897]
```

Now that we have similar users, let's get the recommendations by showing the items bought by these similar users.

Write a function to get similar user recommendations.

```
def knn_recommendation(simular_users_knn):

    #obtaining all the items bought by similar users
    knn_recommnedations = []
    for j in simular_users_knn:
        item_list = data1[data1["CustomerID"]==j]['StockCode'].to_list()
        knn_recommnedations.append(item_list)

    #this gives us multi-dimensional list
    # we need to flatten it
    flat_list = []
    for sublist in knn_recommnedations:
        for item in sublist:
            flat_list.append(item)
    final_recommendations_list = list(dict.fromkeys(flat_list))

    # storing 10 random recommendations in a list
    ten_random_recommendations = random.sample(final_recommendations_list, 10)

    print('Items bought by Similar users based on KNN')

    #returning 10 random recommendations
    return ten_random_recommendations
```

This function replicates the logic used in user-to-user filtering. Next, let's get the final list of items that similar users purchased and recommend any random ten items from it.

Using this function on the previously generated similar users list gets the following recommendations.

```
knn_recommendation(simular_users_knn)
```

The following is the output using the KNN approach.

```
Items bought by Similar users based on KNN
['22487',
 '84997A',
```

```
'22926',
'22921',
'22605',
'23298',
'22916',
'22470',
'22927',
'84978']
```

User 14729 has ten suggestions from the products bought by similar users.

Summary

This chapter covered collaborative filtering-based recommendation engines and implementing the two types of filtering methods—user-to-user and item-to-item—using basic arithmetic operations. The chapter also explored the k-nearest neighbor algorithm (along with some machine learning basics). It ended by implementing user-to-user-based collaborative filtering using the KNN approach. The next chapter explores other popular methods to implement collaborative filtering-based recommendation engines.

CHAPTER 5

Collaborative Filtering Using Matrix Factorization, Singular Value Decomposition, and Co-Clustering

Chapter 4 explored collaborative filtering and using the KNN method. A few more important methods are covered in this chapter: matrix factorization (MF), singular value decomposition (SVD), and co-clustering. These methods (along with KNN) fall into the model-based collaborative filtering approach. The basic arithmetic method of calculating cosine similarity to find similar users falls into the memory-based approach. Each approach has pros and cons; depending on the use case, you must select the suitable approach.

Figure 5-1 explains the two types of approaches in collaborative filtering.

© Akshay Kulkarni, Adarsha Shivananda, Anoosh Kulkarni, V Adithya Krishnan 2023
A. Kulkarni et al., *Applied Recommender Systems with Python*, https://doi.org/10.1007/978-1-4842-8954-9_5

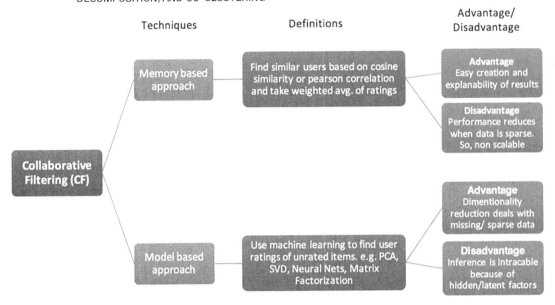

Figure 5-1. *The two approaches of collaborative filtering explained*

The memory-based approach is much easier to implement and explain, but its performance is often affected due to sparse data. But on the other hand, model-based approaches, like MF, handle the sparse data well, but it's usually not intuitive or easy to explain and can be much more complex to implement. But the model-based approach performs better with large datasets and hence is quite scalable.

This chapter focuses on a few popular model-based approaches, such as implementing matrix factorization using the same data from Chapter 4, SVD, and co-clustering models.

Implementation
Matrix Factorization, Co-Clustering, and SVD

The following implementation is a continuation of Chapter 4 and uses the same dataset.
Let's look at the data.

```
data1.head()
```

Figure 5-2 shows the DataFrame from Chapter 4.

	InvoiceNo	StockCode	Quantity	InvoiceDate	DeliveryDate	Discount%	ShipMode	ShippingCost	CustomerID
0	536365	84029E	6	2010-12-01 08:26:00	2010-12-02 08:26:00	0.20	ExpressAir	30.12	17850
1	536365	71053	6	2010-12-01 08:26:00	2010-12-02 08:26:00	0.21	ExpressAir	30.12	17850
2	536365	21730	6	2010-12-01 08:26:00	2010-12-03 08:26:00	0.56	Regular Air	15.22	17850
3	536365	84406B	8	2010-12-01 08:26:00	2010-12-03 08:26:00	0.30	Regular Air	15.22	17850
4	536365	22752	2	2010-12-01 08:26:00	2010-12-04 08:26:00	0.57	Delivery Truck	5.81	17850

Figure 5-2. *Input data*

Let's reuse item_purchase_df from Chapter 4. It is the matrix containing the items and the information on whether customers bought them.

```
items_purchase_df.head()
```

Figure 5-3 shows the item purchase DataFrame/matrix.

CustomerID	12346	12347	12348	12350	12352	12353	12354	12355	12356	12358	...	18269	18270	18272	18273	18278	18280	18281	18282	18283
StockCode																				
10002	0.0	0.0	0.0	0.0	0.0	0.0	0.0	0.0	0.0	0.0	...	0.0	0.0	0.0	0.0	0.0	0.0	0.0	0.0	0.0
10080	0.0	0.0	0.0	0.0	0.0	0.0	0.0	0.0	0.0	0.0	...	0.0	0.0	0.0	0.0	0.0	0.0	0.0	0.0	0.0
10120	0.0	0.0	0.0	0.0	0.0	0.0	0.0	0.0	0.0	0.0	...	0.0	0.0	0.0	0.0	0.0	0.0	0.0	0.0	0.0
10125	0.0	0.0	0.0	0.0	0.0	0.0	0.0	0.0	0.0	0.0	...	0.0	0.0	0.0	0.0	0.0	0.0	0.0	0.0	0.0
10133	0.0	0.0	0.0	0.0	0.0	0.0	0.0	0.0	0.0	0.0	...	0.0	0.0	0.0	0.0	0.0	0.0	0.0	0.0	0.0

5 rows × 3647 columns

Figure 5-3. *Item purchase DataFrame/matrix*

This chapter uses the Python package called surprise for modeling purposes. It has implementations of popular methods in collaborative filtering, like matrix factorization, SVD, co-clustering, and even KNN.

First, let's format the data into the proper format required by the surprise package.

Start by stacking the DataFrame/matrix.

```
data3 = items_purchase_df.stack().to_frame()
#Renaming the column as Quantity
data3 = data3.reset_index().rename(columns={0:"Quantity"})
data3
```

Figure 5-4 shows the output DataFrame after stacking.

	StockCode	CustomerID	Quantity
0	10002	12346	0
1	10002	12347	0
2	10002	12348	0
3	10002	12350	0
4	10002	12352	0
...	...	...	...
12903081	POST	18280	0
12903082	POST	18281	0
12903083	POST	18282	0
12903084	POST	18283	0
12903085	POST	18287	0

12903086 rows × 3 columns

Figure 5-4. *Stacked item purchase DataFrame/matrix*

```
print(items_purchase_df.shape)
print(data3.shape)
```

The following is the output.

```
(3538, 3647)
(12903086, 3)
```

As you can see, items_purchase_df has 3538 unique items (rows) and 3647 unique users (columns). The stacked DataFrame is 3538 × 3647 = 12,903,086 rows, which is too big to pass into any algorithm.

Let's shortlist some customers and items based on the number of orders.

First, put all the IDs in a list.

```
# Storing all customer ids in customers
customer_ids = data1['CustomerID']

# Storing all item descriptions in items
item_ids = data1['StockCode']
```

The following imports the counter to count the number of orders made by each customer and for each item.

```
from collections import Counter
```

Count the number of orders by each customer and store that information in a DataFrame.

```
# counting no. of orders made by each customer
count_orders = Counter(customer_ids)
```

```
# storing the count and customer id in a dataframe
customer_count_df = pd.DataFrame.from_dict(count_orders, orient='index').
reset_index().rename(columns={0:"Quantity"})
```

Drop all customer IDs with less than 120 orders.

```
customer_count_df = customer_count_df[customer_count_df["Quantity"]>120]
```

Rename the index column as 'CustomerID' for the inner join.

```
customer_count_df.rename(columns={'index':'CustomerID'},inplace=True)
customer_count_df
```

Figure 5-5 shows the customer count DataFrame output.

	CustomerID	Quantity
0	17850	297
1	13047	140
2	12583	182
6	14688	265
8	15311	1892
...	...	...
3308	14096	1170
3367	16910	261
3392	16360	226
3413	17728	133
3589	17528	122

568 rows × 2 columns

Figure 5-5. *Customer count DataFrame*

Similarly, repeat the same process for items (i.e., counting the number of orders placed per item and storing it in a DataFrame).

```
# counting no. of times an item was ordered
count_items = Counter(item_ids)
```

```
# storing the count and item description in a dataframe
item_count_df = pd.DataFrame.from_dict(count_items, orient='index').reset_
index().rename(columns={0:"Quantity"})
```

Drop all items that were ordered less than 120 times.

```
item_count_df = item_count_df[item_count_df["Quantity"]>120]
```

Rename the index column as 'Description' for the inner join.

```
item_count_df.rename(columns={'index':'StockCode'},inplace=True)
item_count_df
```

Figure 5-6 shows the output item count DataFrame.

	StockCode	Quantity
0	84029E	161
1	71053	220
3	84406B	213
4	22752	229
5	85123A	1606
...	...	...
3295	23294	181
3296	23295	213
3363	23328	129
3373	23356	148
3376	23355	232

679 rows × 2 columns

Figure 5-6. *Item count DataFrame*

Next, apply a join on both DataFrames with stacked data to create the shortlisted final DataFrame.

```
#Merging stacked df with item count df
data4 = pd.merge(data3, item_count_df, on='StockCode', how='inner')
#Merging with customer count df
data4 = pd.merge(data4, customer_count_df, on='CustomerID', how='inner')
# dropping columns which are not necessary
data4.drop(['Quantity_y','Quantity_x'],axis=1,inplace=True)
data4
```

Figure 5-7 shows the shortlisted DataFrame output.

	StockCode	CustomerID	Quantity
0	10133	12347	124
1	15036	12347	124
2	15056BL	12347	124
3	15056N	12347	124
4	16156S	12347	124
...	...	...	...
385667	85132C	18283	447
385668	85150	18283	447
385669	85152	18283	447
385670	M	18283	447
385671	POST	18283	447

385672 rows × 3 columns

Figure 5-7. *The final shortlisted DataFrame*

Now that the size of the data has been reduced, let's describe it and view the stats.

```
data4.describe()
```

Figure 5-8 describes the shortlisted DataFrame.

	CustomerID	Quantity
count	385672.000000	385672.000000
mean	15360.985915	279.089789
std	1719.468125	337.879413
min	12347.000000	121.000000
25%	13996.250000	151.000000
50%	15413.000000	198.000000
75%	16840.000000	290.000000
max	18283.000000	5095.000000

Figure 5-8. *Describes the shortlisted DataFrame*

You can see from the output that the count has significantly reduced to 385,672 records, from 12,903,086. But this DataFrame is to be formatted further using built-in functions from the surprise package to be supported.

Read the data in a format supported by the surprise library.

```
reader = Reader(rating_scale=(0,5095))
```

The range has been set as 0,5095 because the maximum quantity value is 5095. Load the dataset in a format supported by the surprise library.

```
formated_data = Dataset.load_from_df(data4, reader)
```

The final formatted data is ready.

Now, let's split the data to train and test for validating the models.

```
# performing train test split on the dataset
train_set, test_set = train_test_split(formated_data, test_size= 0.2)
```

Implementing NMF

Let's start by modeling the non-negative matrix factorization method.

Figure 5-9 explains matrix factorization (multiplication).

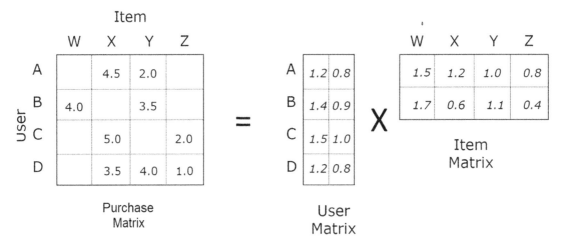

Figure 5-9. *Matrix factorization*

Matrix factorization is a popular method used in building collaborative filtering-based recommendation systems. It is a basic embedding model where latent/hidden features (embeddings) are generated from the user and item matrixes using matrix multiplication. This reduces the dimensionality of the full input matrix and hence is a compact representation, increasing the scalability and performance. These latent features are then used to fit an optimization problem (usually minimizing an error equation) to get to the predictions.

```
# defining the model
algo1 = NMF()

# model fitting
```

```
algo1.fit(train_set)

# model prediction
pred1 = algo1.test(test_set)
```

Using built-in functions, you can calculate the performance metrics like RMSE (root-mean-squared error) and MAE (mean absolute error).

```
# RMSE
accuracy.rmse(pred1)

#MAE
accuracy.mae(pred1)
```

The following is the output.

```
RMSE: 428.3167
MAE:   272.6909
```

The RMSE and MAE are moderately high for this model, so let's try the other two and compare them at the end.

You can also cross-validate (using built-in functions) to further validate these values.

```
cross_validate(algo1, formated_data, verbose=True)
```

Figure 5-10 shows the cross-validation output for NMF.

```
Evaluating RMSE, MAE of algorithm NMF on 5 split(s).

                 Fold 1  Fold 2  Fold 3  Fold 4  Fold 5  Mean     Std
RMSE (testset)   408.786 439.814 0413.959 4454.817 1421.493 6427.774 117.1296
MAE (testset)    266.910 3275.054 3274.251 4275.623 6271.296 4272.627 23.2237
Fit time         0.13    0.13    0.12    0.13    0.14    0.13    0.00
Test time        0.03    0.03    0.03    0.02    0.03    0.03    0.00
```

Figure 5-10. *Cross-validation output for NMF*

The cross-validation shows that the average RMSE is 427.774, and MAE is approximately 272.627, which is moderately high.

Implementing Co-Clustering

Co-clustering (also known as *bi-clustering*) is commonly used in collaborative filtering. It is a data-mining technique that simultaneously clusters the columns and rows of a DataFrame/matrix. It differs from normal clustering, where each object is checked for similarity with other objects based on a single entity/type of comparison. As in co-clustering, you check for co-grouping of two different entities/types of comparison for each object simultaneously as a pairwise interaction.

Let's try modeling with the co-clustering method.

```
# defining the model
algo2 = CoClustering()

# model fitting
algo2.fit(train_set)

# model prediction
pred2 = algo2.test(test_set)
```

Calculate the RMSE and MAE performance metrics using built-in functions.

```
# RMSE
accuracy.rmse(pred2)

#MAE
accuracy.mae(pred2)
```

The following is the output.

```
RMSE: 6.7877
MAE:  5.8950
```

The RMSE and MAE are very low for this model. Until now, this has performed the best (better than NMF).

Cross-validate (using built-in functions) to further validate these values.

```
cross_validate(algo2, formated_data, verbose=True)
```

Figure 5-11 shows the cross-validation output for co-clustering.

```
Evaluating RMSE, MAE of algorithm CoClustering on 5 split(s).

                    Fold 1  Fold 2  Fold 3   Fold 4   Fold 5   Mean     Std
RMSE (testset)      6.8485  6.6710  34.0950  11.0666  11.4735  14.0309  10.2338
MAE (testset)       5.6185  5.0401  7.0667   7.0208   5.9296   6.1352   0.7950
Fit time            0.19    0.17    0.18     0.20     0.18     0.18     0.01
Test time           0.01    0.01    0.02     0.01     0.01     0.01     0.00
```

Figure 5-11. *Cross-validation output for co-clustering*

The cross-validation shows that the average RMSE is 14.031, and MAE is
approximately 6.135, which is quite low.

Implementing SVD

Singular value decomposition is a linear algebra concept generally used as a
dimensionality reduction method. It is also a type of matrix factorization. It works
similarly in collaborative filtering, where a matrix with rows and columns as users and
items is reduced further into latent feature matrixes. An error equation is minimized to
get to the prediction.

Let's try modeling using the SVD method.

```
# defining the model
import SVD
algo3 = SVD()

# model fitting
algo3.fit(train_set)

# model prediction
pred3 = algo3.test(test_set)
```

Calculate the RMSE and MAE performance metrics using built-in functions.

```
# RMSE
accuracy.rmse(pred3)

#MAE
accuracy.mae(pred3)
```

The following is the output.

```
RMSE: 4827.6830
MAE:  4815.8341
```

The RMSE and MAE are significantly high for this model. Until now, this has performed the worst (worse than NMF and co-clustering).

Cross-validate (using built-in functions) to further validate these values.

```
cross_validate(algo3, formated_data, verbose=True)
```

Figure 5-12 shows the cross-validation output for SVD.

```
Evaluating RMSE, MAE of algorithm SVD on 5 split(s).

                Fold 1  Fold 2  Fold 3  Fold 4  Fold 5  Mean    Std
RMSE (testset)  4826.9477 4824.5461 4833.9809 4821.3278 4831.9276 4827.7460 4.6568
MAE (testset)   4814.7419 4812.7887 4823.2951 4807.1767 4821.5486 4815.9102 5.8943
Fit time        0.11    0.10    0.10    0.10    0.10    0.10    0.00
Test time       0.04    0.04    0.04    0.04    0.04    0.04    0.00
```

Figure 5-12. *Cross-validation output for SVD*

The cross-validation shows that the average RMSE is 4831.928 and MAE is approximately 4821.549, which is very high.

Getting the Recommendations

The co-clustering model has performed better than the NMF and the SVD models. But let's validate the model further once more before using the predictions.

For validating the model, let's use item 47590B and customer 15738.

```
data1[(data1['StockCode']=='47590B')&(data1['CustomerID']==15738)].
Quantity.sum()
```

The following is the output.

```
78
```

Let's get the prediction for the same combination to see the estimation or prediction.

```
algo2.test([['47590B',15738,78]])
```

The following is the output.

```
[Prediction(uid='47590B', iid=15738, r_ui=78, est=133.01087456331527,
details={'was_impossible': False})]
```

The predicted value given by the model is 133.01, while the actual was 78. It is close to the actual and validated the model performance even further.

The predictions are from the co-clustering model.

```
pred2
```

The following is the output.

```
[Prediction(uid='85014B', iid=17228, r_ui=130.0, est=119.18329013727276,
 details={'was_impossible': False}),
 Prediction(uid='84406B', iid=16520, r_ui=156.0, est=161.85867140088936,
 details={'was_impossible': False}),
 Prediction(uid='47590B', iid=17365, r_ui=353.0, est=352.7773176826455,
 details={'was_impossible': False}),
...,
 Prediction(uid='85049G', iid=16755, r_ui=170.0, est=159.5403752414615,
 details={'was_impossible': False}),
 Prediction(uid='16156S', iid=14895, r_ui=367.0, est=368.129814201444,
 details={'was_impossible': False}),
 Prediction(uid='47566B', iid=17238, r_ui=384.0, est=393.60123986750034,
 details={'was_impossible': False})]
```

Now let's use these predictions and see the best and the worst predictions, but first, let's get the final output onto a DataFrame.

```
predictions_data = pd.DataFrame(pred2, columns=['item_id', 'customer_id',
'quantity', 'prediction', 'details'])
```

But first, let's also add important information like the number of item orders and customer orders for each record using the following function.

```
def get_item_orders(user_id):
    try:
        # for an item, return the no. of orders made
        return len(train_set.ur[train_set.to_inner_uid(user_id)])
```

```
    except ValueError:
        # user not present in training
        return 0

def get_customer_orders(item_id):
    try:
        # for an customer, return the no. of orders made
        return len(train_set.ir[train_set.to_inner_iid(item_id)])
    except ValueError:
        # item not present in training
        return 0
```

The following calls these functions.

```
predictions_data['item_orders'] = predictions_data.item_id.apply(get_
item_orders)
predictions_data['customer_orders'] = predictions_data.customer_
id.apply(get_customer_orders)
```

Calculate the error component to get the best and worst predictions.

```
predictions_data['error'] = abs(predictions_data.prediction - predictions_
data.quantity)
predictions_data
```

Figure 5-13 shows the prediction DataFrame.

	item_id	customer_id	quantity	prediction	details	item_orders	customer_orders	error
0	85014B	17228	130.0	119.183290	{'was_impossible': False}	459	31	10.816710
1	84406B	16520	156.0	161.858671	{'was_impossible': False}	459	29	5.858671
2	47590B	17365	353.0	352.777318	{'was_impossible': False}	457	32	0.222682
3	85049G	16755	170.0	159.540375	{'was_impossible': False}	450	32	10.459625
4	16156S	14895	367.0	368.129814	{'was_impossible': False}	440	30	1.129814
...	...	...	...	...	...	...	...	...
4539	47590B	15764	180.0	179.777318	{'was_impossible': False}	457	30	0.222682
4540	84970L	16222	137.0	144.853747	{'was_impossible': False}	458	34	7.853747
4541	84596F	16340	153.0	154.254839	{'was_impossible': False}	453	29	1.254839
4542	85099B	17511	745.0	748.576631	{'was_impossible': False}	447	32	3.576631
4543	85049E	16265	194.0	190.137590	{'was_impossible': False}	458	29	3.862410

4544 rows × 8 columns

Figure 5-13. *Prediction DataFrame*

The following gets the best predictions.

```
best_predictions = predictions_data.sort_values(by='error')[:10]
best_predictions
```

Figure 5-14 shows the best predictions.

	item_id	customer_id	quantity	prediction	details	item_orders	customer_orders	error
334	16156S	17841	5095.0	5095.000000	{'was_impossible': False}	440	32	0.000000
3973	47590B	13230	457.0	456.777318	{'was_impossible': False}	457	29	0.222682
697	47590B	12415	601.0	600.777318	{'was_impossible': False}	457	30	0.222682
2339	47590B	13869	307.0	306.777318	{'was_impossible': False}	457	34	0.222682
1572	47590B	13078	276.0	275.777318	{'was_impossible': False}	457	32	0.222682
1608	47590B	17428	299.0	298.777318	{'was_impossible': False}	457	35	0.222682
1160	47590B	17799	343.0	342.777318	{'was_impossible': False}	457	31	0.222682
574	47590B	17337	543.0	542.777318	{'was_impossible': False}	457	29	0.222682
4000	47590B	14527	694.0	693.777318	{'was_impossible': False}	457	35	0.222682
516	47590B	14701	238.0	237.777318	{'was_impossible': False}	457	31	0.222682

Figure 5-14. *Best predictions*

The following gets the worst predictions.

```
worst_predictions = predictions_data.sort_values(by='error')[-10:]
worst_predictions
```

Figure 5-15 shows the worst predictions.

	item_id	customer_id	quantity	prediction	details	item_orders	customer_orders	error
4003	47599A	14286	141.0	125.720820	{'was_impossible': False}	471	34	15.279180
2939	47599A	15696	122.0	106.720820	{'was_impossible': False}	471	28	15.279180
2933	47599A	16393	214.0	198.720820	{'was_impossible': False}	471	32	15.279180
538	47599A	12662	157.0	141.720820	{'was_impossible': False}	471	32	15.279180
537	47599A	14040	178.0	162.720820	{'was_impossible': False}	471	31	15.279180
2180	47599A	14808	208.0	192.720820	{'was_impossible': False}	471	31	15.279180
1585	47599A	13555	136.0	120.720820	{'was_impossible': False}	471	30	15.279180
3252	47599A	14911	3648.0	3632.720820	{'was_impossible': False}	471	34	15.279180
1651	47599A	13089	1511.0	1495.720820	{'was_impossible': False}	471	31	15.279180
3033	47599A	12949	179.0	163.009478	{'was_impossible': False}	471	31	15.990522

Figure 5-15. *Worst predictions*

You can now use the predictions data to get to the recommendations. First, find the customers that have bought the same items as a given user, and then from the other items they have bought, to fetch the top items and recommend them.

Let's again use customer 12347 and create a list of the items this user bought.

```
# Getting item list for user 12347
item_list = predictions_data[predictions_data['customer_id']==12347]['item_
id'].values.tolist()
item_list
```

The following is the output.

```
['82494L',
 '84970S',
 '47599A',
 '84997B',
 '85123A',
 '84997C',
 '85049A']
```

Get the list of customers who bought the same items as user 12347.

```
# Getting list of unique customers who also bought same items (item_list)
```

```
customer_list = predictions_data[predictions_data['item_id'].isin(item_
list)]['customer_id'].values
customer_list = np.unique(customer_list).tolist()
customer_list
```

The following is the output.

```
[12347,
 12362,
 12370,
 12378,
 ...,
 12415,
 12417,
 12428]
```

Now let's filter these customers (customer_list) from predictions data, remove the items already bought, and recommend the top items (prediction).

```
# filtering those customers from predictions data
filtered_data = predictions_data[predictions_data['customer_id'].
isin(customer_list)]

# removing the items already bought
filtered_data = filtered_data[~filtered_data['item_id'].isin(item_list)]

# getting the top items (prediction)
recommended_items = filtered_data.sort_values('prediction',ascending=
False).reset_index(drop=True).head(10)['item_id'].values.tolist()
recommended_items
```

The following is the output.

```
['16156S',
 '85049E',
 '47504K',
 '85099C',
 '85049G',
 '85014B',
```

```
'72351B',
'84536A',
'48173C',
'47590A']
```

The recommended list of items for user 12347 is achieved.

Summary

This chapter continued the discussion of collaborative filtering-based recommendation engines. Popular methods like matrix factorization, SVD, and co-clustering were explored with a focus on implementing all three models. For the given data, the co-clustering method performed the best, but you need to try all the different methods available to see which best fits your data and use case in building a recommendation system.

CHAPTER 6

Hybrid Recommender Systems

The previous chapters implemented recommendation engines using content-based and collaborative-based filtering methods. Each method has its pros and cons. Collaborative filtering suffers from cold-start, which means when there is a new customer or item in the data, recommendation won't be possible.

Content-based filtering tends to recommend similar items to that purchased/liked before, becoming repetitive. There is no personalization effect in this case.

Figure 6-1 explains hybrid recommendation systems.

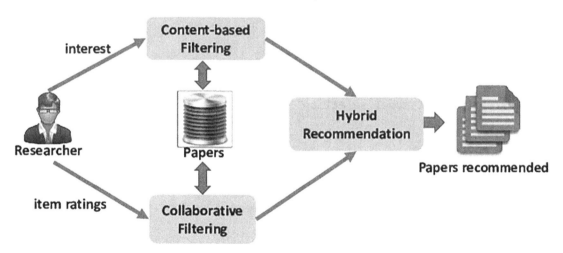

Figure 6-1. *Hybrid recommendation systems*

Reference: https://www.researchgate.net/profile/Xiangjie-Kong-2/ publication/330077673/figure/fig5/AS:710433577107459@1546391972632/A-hybrid-paper-recommendation-system.png

© Akshay Kulkarni, Adarsha Shivananda, Anoosh Kulkarni, V Adithya Krishnan 2023
A. Kulkarni et al., *Applied Recommender Systems with Python*, https://doi.org/10.1007/978-1-4842-8954-9_6

To tackle some of these cons, introducing hybrid recommendations systems. Hybrid recommendation systems use a hybrid model (i.e., combining content-based and collaborative filtering methods). It will not only help to overcome the shortcomings of the individual models but also increase efficiency and give better recommendations in most cases.

This chapter implements a hybrid recommendation engine built to recommend products used for an e-commerce company. The LightFM Python package is used for this implementation.

For more information, refer to the LightFM documentation at `https://making.lyst.com/lightfm/docs/home.html`.

Implementation

Let's import all the required libraries.

```
import pandas as pd
import numpy as np
from scipy.sparse import coo_matrix # for constructing sparse matrix
from lightfm import LightFM # for model
from lightfm.evaluation import auc_score
import time
import sklearn
from sklearn import model_selection
```

Data Collection

This chapter uses the same custom e-commerce dataset used in previous chapters. It can be found at `github.com/apress/applied-recommender-systems-python`.

The following reads the data.

```
#orders data
order_df = pd.read_excel('Rec_sys_data.xlsx','order')
#customers data
customer_df = pd.read_excel('Rec_sys_data.xlsx','customer')
#products data
product_df = pd.read_excel('Rec_sys_data.xlsx','product')
order_df.head()
```

Figure 6-2 shows the orders DataFrame.

	InvoiceNo	StockCode	Quantity	InvoiceDate	DeliveryDate	Discount%	ShipMode	ShippingCost	CustomerID
0	536365	84029E	6	2010-12-01 08:26:00	2010-12-02 08:26:00	0.20	ExpressAir	30.12	17850
1	536365	71053	6	2010-12-01 08:26:00	2010-12-02 08:26:00	0.21	ExpressAir	30.12	17850
2	536365	21730	6	2010-12-01 08:26:00	2010-12-03 08:26:00	0.56	Regular Air	15.22	17850
3	536365	84406B	8	2010-12-01 08:26:00	2010-12-03 08:26:00	0.30	Regular Air	15.22	17850
4	536365	22752	2	2010-12-01 08:26:00	2010-12-04 08:26:00	0.57	Delivery Truck	5.81	17850

Figure 6-2. *Orders data*

```
customer_df.head()
```

Figure 6-3 shows the customers DataFrame.

	CustomerID	Gender	Age	Income	Zipcode	Customer Segment
0	13089	male	53	High	8625	Small Business
1	15810	female	22	Low	87797	Small Business
2	15556	female	29	High	29257	Corporate
3	13137	male	29	Medium	97818	Middle class
4	16241	male	36	Low	79200	Small Business

Figure 6-3. *Customers data*

```
product_df.head()
```

Figure 6-4 shows the products DataFrame.

	StockCode	Product Name	Description	Category	Brand	Unit Price
0	22629	Ganma Superheroes Ordinary Life Case For Samsu...	New unique design, great gift.High quality pla...	Cell Phones|Cellphone Accessories|Cases & Prot...	Ganma	13.99
1	21238	Eye Buy Express Prescription Glasses Mens Wome...	Rounded rectangular cat-eye reading glasses. T...	Health|Home Health Care|Daily Living Aids	Eye Buy Express	19.22
2	22181	MightySkins Skin Decal Wrap Compatible with Ni...	Each Nintendo 2DS kit is printed with super-hi...	Video Games|Video Game Accessories|Accessories...	Mightyskins	14.99
3	84879	Mediven Sheer and Soft 15-20 mmHg Thigh w/ Lac...	The sheerest compression stocking in its class...	Health|Medicine Cabinet|Braces & Supports	Medi	62.38
4	84836	Stupell Industries Chevron Initial Wall D cor	Features: -Made in the USA. -Sawtooth hanger o...	Home Improvement|Paint|Wall Decals|All Wall De...	Stupell Industries	35.99

Figure 6-4. *Products data*

Merge the data.

```
#merging all three data frames
```

```
merged_df = pd.merge(order_df,customer_df,left_on=['CustomerID'], right_
on=['CustomerID'], how='left')
merged_df = pd.merge(merged_df,product_df,left_on=['StockCode'], right_
on=['StockCode'], how='left')
merged_df.head()
```

Figure 6-5 shows the merged DataFrame that will be used.

InvoiceNo	StockCode	Quantity	InvoiceDate	DeliveryDate	Discount%	ShipMode	ShippingCost	CustomerID	Gender	Age	Income	Zipcode	Customer Segment	Pr(
536365	84029E	6	2010-12-01 08:26:00	2010-12-02 08:26:00	0.20	ExpressAir	30.12	17850	female	48	Medium	84306	Middle class	3 1/2 2 Fu Craft Smc
536365	71053	6	2010-12-01 08:26:00	2010-12-02 08:26:00	0.21	ExpressAir	30.12	17850	female	48	Medium	84306	Middle class	Awl Shar Fl Pa
536365	21730	6	2010-12-01 08:26:00	2010-12-03 08:26:00	0.56	Regular Air	15.22	17850	female	48	Medium	84306	Middle class	Ebe Rect Ha Hir
536365	84406B	8	2010-12-01 08:26:00	2010-12-03 08:26:00	0.30	Regular Air	15.22	17850	female	48	Medium	84306	Middle class	Mighty Skin Comp with
536365	22752	2	2010-12-01 08:26:00	2010-12-04 08:26:00	0.57	Delivery Truck	5.81	17850	female	48	Medium	84306	Middle class	awe since birthd t-

Figure 6-5. *Merged data*

Data Preparation

Before building the recommendation model, the required data must be in the proper format so that the model can take input. Let's get the user-to-product interaction matrix and product-to-features interaction mappings.

Start with getting the list of unique users and unique products. Write two functions to get the unique lists.

```
def unique_users(data, column):
    return np.sort(data[column].unique())
def unique_items(data, column):
    item_list = data[column].unique()
    return item_list
```

Create unique lists.

```
user_list = unique_users(order_df, "CustomerID")
item_list = unique_items(product_df, "Product Name")

user_list
```

The following is the output.

```
array([12346, 12347, 12348, ..., 18282, 18283, 18287], dtype=int64)
```

```
item_list
```

The following is the output.

```
array(['Ganma Superheroes Ordinary Life Case For Samsung Galaxy Note 5 Hard
Case Cover',
       'Eye Buy Express Prescription Glasses Mens Womens Burgundy Crystal
       Clear Yellow Rounded Rectangular Reading Glasses Anti Glare grade',
       ...,
       'Mediven Sheer and Soft 15-20 mmHg Thigh w/ Lace Silicone Top Band
       CT Wheat II - Ankle 8-8.75 inches',
       Union 3" Female Ports Stainless Steel Pipe Fitting',
       'Auburn Leathercrafters Tuscany Leather Dog Collar',
       '3 1/2"W x 32"D x 36"H Traditional Arts & Crafts Smooth Bracket,
       Douglas Fir'])
```

Let's create a function to get the total list of unique values given three feature names from a DataFrame. It gets the total unique list for three features: Customer Segment, Age, and Gender.

```
def features_to_add(customer, column1,column2,column3):
    customer1 = customer[column1]
    customer2 = customer[column2]
    customer3 = customer[column3]
    return pd.concat([customer1,customer3,customer2], ignore_index = True).
    unique()
```

Call the function for these features.

```
feature_unique_list = features_to_add(customer_df,'Customer
Segment',"Age","Gender")
feature_unique_list
```

The following is the output.

```
array(['Small Business', 'Corporate', 'Middle class', 'male', 'female',
       53, 22, 29, 36, 48, 45, 47, 23, 39, 34, 52, 51, 35, 19, 26, 37, 18,
       20, 21, 41, 31, 28, 50, 38, 30, 25, 32, 55, 43, 54, 49, 40, 33, 44,
       46, 42, 27, 24], dtype=object)
```

Now that we have the unique list for users, products, and features, we need to create ID mappings to convert user_id, item_id, and feature_id into integer indices because LightFM can't read any other data types.

Let's write a function for that.

```
def mapping(user_list, item_list, feature_unique_list):
    #creating empty output dicts
    user_to_index_mapping = {}
    index_to_user_mapping = {}
    # Create id mappings to convert user_id
    for user_index, user_id in enumerate(user_list):
        user_to_index_mapping[user_id] = user_index
        index_to_user_mapping[user_index] = user_id

    item_to_index_mapping = {}
    index_to_item_mapping = {}
    # Create id mappings to convert item_id
    for item_index, item_id in enumerate(item_list):
        item_to_index_mapping[item_id] = item_index
        index_to_item_mapping[item_index] = item_id

    feature_to_index_mapping = {}
    index_to_feature_mapping = {}
    # Create id mappings to convert feature_id
    for feature_index, feature_id in enumerate(feature_unique_list):
        feature_to_index_mapping[feature_id] = feature_index
        index_to_feature_mapping[feature_index] = feature_id
```

```
return user_to_index_mapping, index_to_user_mapping, \
       item_to_index_mapping, index_to_item_mapping, \
       feature_to_index_mapping, index_to_feature_mapping
```

Call the function by giving user_list, item_list, and feature_unique_list as input.

```
user_to_index_mapping, index_to_user_mapping, \
       item_to_index_mapping, index_to_item_mapping, \
       feature_to_index_mapping, index_to_feature_mapping =
       mapping(user_list, item_list, feature_unique_list)

user_to_index_mapping
```

The following is the output.

```
{12346: 0,
 12347: 1,
 12348: 2,
 12350: 3,
 12352: 4,
...}
```

Now let's fetch the user-to-product relationship and calculate the total quantity per user.

```
user_to_product = merged_df[['CustomerID','Product Name','Quantity']]
#Calculating the total quantity(sum) per customer-product
user_to_product = user_to_product.groupby(['CustomerID','Product Name']).
agg({'Quantity':'sum'}).reset_index()
user_to_product.tail()
```

Figure 6-6 shows the user-to-product relationship data.

	CustomerID	Product Name	Quantity
138397	18287	Sport-Tek Ladies PosiCharge Competitor Tee	24
138398	18287	Ultra Sleek And Spacious Pearl White Lacquer 1...	6
138399	18287	Union 3" Female Ports Stainless Steel Pipe Fit...	12
138400	18287	awesome since 1948 - 69th birthday gift t-shir...	4
138401	18287	billyboards Porcelain Menu Chalkboard	6

Figure 6-6. *User-to-product relationship data*

Similarly, let's get the product-to-features relationship data.

```
product_to_feature = merged_df[['Product Name','Customer
Segment','Quantity']]
#Calculating the total quantity(sum) per customer_segment-product
product_to_feature = product_to_feature.groupby(['Product Name','Customer
Segment']).agg({'Quantity':'sum'}).reset_index()
product_to_feature.head()
```

Figure 6-7 shows the product-to-features relationship data.

	Product Name	Customer Segment	Quantity
0	"In Vinyl W.e Trust" Rasta Quote Men's T-shirt	Corporate	712
1	"In Vinyl W.e Trust" Rasta Quote Men's T-shirt	Middle class	272
2	"In Vinyl W.e Trust" Rasta Quote Men's T-shirt	Small Business	388
3	"Soccer" Vinyl Graphic - Large - Ivory	Corporate	1940
4	"Soccer" Vinyl Graphic - Large - Ivory	Middle class	1418

***Figure 6-7.** Product-to-features relationship data*

Let's split the user-to-product relationship into train and test data.

```
user_to_product_train,user_to_product_test = model_selection.train_test_
split(user_to_product,test_size=0.33, random_state=42)

print("Training set size:")
print(user_to_product_train.shape)
print("Test set size:")
print(user_to_product_test.shape)
```

The following is the output.

```
Training set size:
(92729, 3)
Test set size:
(45673, 3)
```

Now that the data and the ID mappings are in place, to get the user-to-product and product-to-features interaction matrix, let's first create a function that returns the interaction matrix.

```
def interactions(data, row, col, value, row_map, col_map):

    #converting the row with its given mappings
    row = data[row].apply(lambda x: row_map[x]).values
    #converting the col with its given mappings
    col = data[col].apply(lambda x: col_map[x]).values
    value = data[value].values
    #returning the interaction matrix
    return coo_matrix((value, (row, col)), shape = (len(row_map),
    len(col_map)))
```

Then let's generate user_item_interaction_matrix for train and test data using the preceding function.

```
#for train
user_to_product_interaction_train = interactions(user_to_product_train,
"CustomerID",
"Product Name", "Quantity", user_to_index_mapping, item_to_index_mapping)
```

```
#for test
user_to_product_interaction_test = interactions(user_to_product_test,
"CustomerID",
"Product Name", "Quantity", user_to_index_mapping, item_to_index_mapping)
```

```
print(user_to_product_interaction_train)
```

The following is the output.

```
(2124, 230)   10
(1060, 268)   16
  :            :
(64, 8)       24
(3406, 109)    1
(3219, 12)    12
```

Similarly, let's generate the product-to-features interaction matrix.

```
product_to_feature_interaction = interactions(product_to_feature, "Product
Name", "Customer Segment","Quantity",item_to_index_mapping, feature_to_
index_mapping)
```

Model Building

The data is in the correct format, so let's begin the modeling process. This chapter uses the LightFM model, which can incorporate user and item metadata to form robust hybrid recommendation models.

Let's try multiple models and then choose the one with the best performance. These models have different hyperparameters, so this is part of the hyperparameter tuning stage of modeling.

The loss function used in the model is one of the parameters to tune. The three values are warp, logistic, and bpr.

Let's start the model-building experiment.

Attempt 1 is loss = warp, epochs = 1, and num_threads = 4.

```
# initialising model with warp loss function
model_with_features = LightFM(loss = "warp")

start = time.time()
#====================
# fitting the model with hybrid collaborative filtering + content based
(product + features)

model_with_features.fit_partial(user_to_product_interaction_train,
          user_features=None,
          item_features=product_to_feature_interaction,
          sample_weight=None,
          epochs=1,
          num_threads=4,
          verbose=False)

#====================
end = time.time()
print("time taken = {0:.{1}f} seconds".format(end - start, 2))
```

The following is the output.

```
time taken = 0.11 seconds
```

Calculate the area under the curve (AUC) score for validation.

```
start = time.time()
#===================
# Getting the AUC score using in-built function
auc_with_features = auc_score(model = model_with_features,
                       test_interactions = user_to_product_
                       interaction_test,
                       train_interactions = user_to_product_
                       interaction_train,
                       item_features = product_to_feature_interaction,
                       num_threads = 4, check_intersections=False)
#===================
end = time.time()
print("time taken = {0:.{1}f} seconds".format(end - start, 2))

print("average AUC without adding item-feature interaction = {0:.{1}f}".
format(auc_with_features.mean(), 2))
```

The following is the output.

```
time taken = 0.24 seconds
average AUC without adding item-feature interaction = 0.17
```

Attempt 2 is loss = logistic, epochs = 1, and num_threads = 4.

```
# initialising model with warp loss function
model_with_features = LightFM(loss = "logistic")

start = time.time()
#===================
# fitting the model with hybrid collaborative filtering + content based
(product + features)

model_with_features.fit_partial(user_to_product_interaction_train,
          user_features=None,
          item_features=product_to_feature_interaction,
          sample_weight=None,
```

```
        epochs=1,
        num_threads=4,
        verbose=False)

#===================
end = time.time()
print("time taken = {0:.{1}f} seconds".format(end - start, 2))
```

The following is the output.

```
time taken = 0.11 seconds
```

Calculate the AUC score for the preceding model.

```
start = time.time()
#===================
# Getting the AUC score using in-built function

auc_with_features = auc_score(model = model_with_features,
                    test_interactions = user_to_product_
                    interaction_test,
                    train_interactions = user_to_product_
                    interaction_train,
                    item_features = product_to_feature_interaction,
                    num_threads = 4, check_intersections=False)
#===================
end = time.time()
print("time taken = {0:.{1}f} seconds".format(end - start, 2))

print("average AUC without adding item-feature interaction = {0:.{1}f}".
format(auc_with_features.mean(), 2))
```

The following is the output.

```
time taken = 0.22 seconds
average AUC without adding item-feature interaction = 0.89
```

Attempt 3 is loss = bpr, epochs = 1, and num_threads = 4.

```
# initialising model with warp loss function
model_with_features = LightFM(loss = "bpr")
```

```
start = time.time()
#===================
# fitting the model with hybrid collaborative filtering + content based
(product + features)

model_with_features.fit_partial(user_to_product_interaction_train,
        user_features=None,
        item_features=product_to_feature_interaction,
        sample_weight=None,
        epochs=1,
        num_threads=4,
        verbose=False)

#===================
end = time.time()
print("time taken = {0:.{1}f} seconds".format(end - start, 2))
```

The following is the output.

```
time taken = 0.12 seconds
```

Calculate the AUC score for the preceding model.
The following is the output.
Attempt 4 is loss = logistic, epochs = 10, and num_threads = 20.
The following is the output.
Calculate the AUC score for the preceding model.

```
start = time.time()
#===================
# Getting the AUC score using in-built function

auc_with_features = auc_score(model = model_with_features,
                    test_interactions = user_to_product_
                    interaction_test,
                    train_interactions = user_to_product_
                    interaction_train,
                    item_features = product_to_feature_interaction,
                    num_threads = 4, check_intersections=False)
#===================
```

```
end = time.time()
print("time taken = {0:.{1}f} seconds".format(end - start, 2))

print("average AUC without adding item-feature interaction = {0:.{1}f}".
format(auc_with_features.mean(), 2))
```

The following is the output.

```
time taken = 0.22 seconds
average AUC without adding item-feature interaction = 0.38
model_with_features = LightFM(loss = "logistic")

start = time.time()
#====================
# fitting the model with hybrid collaborative filtering + content based
(product + features)

model_with_features.fit_partial(user_to_product_interaction_train,
            user_features=None,
            item_features=product_to_feature_interaction,
            sample_weight=None,
            epochs=10,
            num_threads=20,
            verbose=False)

#====================
end = time.time()
print("time taken = {0:.{1}f} seconds".format(end - start, 2))
time taken = 0.77 seconds
start = time.time()
#====================
# Getting the AUC score using in-built function

auc_with_features = auc_score(model = model_with_features,
                        test_interactions = user_to_product_
                        interaction_test,
                        train_interactions = user_to_product_
                        interaction_train,
                        item_features = product_to_feature_interaction,
```

```
                         num_threads = 4, check_intersections=False)
#===================
end = time.time()
print("time taken = {0:.{1}f} seconds".format(end - start, 2))

print("average AUC without adding item-feature interaction = {0:.{1}f}".
format(auc_with_features.mean(), 2))
time taken = 0.25 seconds
average AUC without adding item-feature interaction = 0.89
```

The last model (logistic) performed the best overall (highest AUC score). Let's merge the train and test and do a final training by using the parameters from the logistic model, which gave 0.89 AUC.

Merge the train and test with the following function.

```
def train_test_merge(training_data, testing_data):

    # initialising train dict
    train_dict = {}
    for row, col, data in zip(training_data.row, training_data.col,
    training_data.data):
        train_dict[(row, col)] = data

    # replacing with the test set

    for row, col, data in zip(testing_data.row, testing_data.col, testing_
    data.data):
        train_dict[(row, col)] = max(data, train_dict.get((row, col), 0))

    # converting to the row
    row_list = []
    col_list = []
    data_list = []
    for row, col in train_dict:
        row_list.append(row)
        col_list.append(col)
        data_list.append(train_dict[(row, col)])
```

```
# converting to np array

row_list = np.array(row_list)
col_list = np.array(col_list)
data_list = np.array(data_list)

#returning the matrix output
return coo_matrix((data_list, (row_list, col_list)), shape = (training_
data.shape[0], training_data.shape[1]))
```

Call the preceding function to get the final (full) data to build the final model.

```
user_to_product_interaction = train_test_merge(user_to_product_interaction_
train, user_to_product_interaction_test)
```

Final Model after Combining the Train and Test Data

Let's build the LightFM model with loss = logistic, epochs = 10, and num_threads = 20.

```
# retraining the final model with combined dataset

final_model = LightFM(loss = "warp",no_components=30)

# fitting to combined dataset

start = time.time()
#===================
#final model fitting

final_model.fit(user_to_product_interaction,
        user_features=None,
        item_features=product_to_feature_interaction,
        sample_weight=None,
        epochs=10,
        num_threads=20,
        verbose=False)

#===================
```

```
end = time.time()
print("time taken = {0:.{1}f} seconds".format(end - start, 2))
```

The following is the output.

```
time taken = 3.46 seconds
```

Getting Recommendations

Now that the hybrid recommendation model is ready, let's use it to get the recommendations for a given user.

Let's write a function for getting those recommendations given a user id as input.

```
def get_recommendations(model,user,items,user_to_product_interaction_
matrix,user2index_map,product_to_feature_interaction_matrix):

# getting the userindex

    userindex = user2index_map.get(user, None)

    if userindex == None:
        return None

    users = userindex

    # getting products already bought

    known_positives = items[user_to_product_interaction_matrix.tocsr()
    [userindex].indices]
    print('User index =',users)

    # scores from model prediction
    scores = model.predict(user_ids = users, item_ids = np.arange(user_to_
    product_interaction_matrix.shape[1]),item_features=product_to_feature_
    interaction_matrix)

    #getting top items
```

```
top_items = items[np.argsort(-scores)]

# printing out the result
print("User %s" % user)
print("    Known positives:")

for x in known_positives[:10]:
    print("                    %s" % x)

print("    Recommended:")

for x in top_items[:10]:
    print("                    %s" % x)
```

This function calculates a user's prediction score (the likelihood to buy) for all items, and the ten highest scored items are recommended. Let's print the known positives or items bought by that user for validation.

Call the following function for a random user (CustomerID 17017) to get recommendations.

```
get_recommendations(final_model,17017,item_list,user_to_product_
interaction,user_to_index_mapping,product_to_feature_interaction)
```

The following is the output.

```
User index = 2888
User 17017

Known positives:
Ganma Superheroes Ordinary Life Case For Samsung Galaxy Note 5 Hard
Case Cover
MightySkins Skin Decal Wrap Compatible with Nintendo Sticker Protective
Cover 100's of Color Options
Mediven Sheer and Soft 15-20 mmHg Thigh w/ Lace Silicone Top Band CT Wheat
II - Ankle 8-8.75 inches
MightySkins Skin Decal Wrap Compatible with OtterBox Sticker Protective
Cover 100's of Color Options
MightySkins Skin Decal Wrap Compatible with DJI Sticker Protective Cover
100's of Color Options
```

MightySkins Skin Decal Wrap Compatible with Lenovo Sticker Protective Cover 100's of Color Options
Ebe Reading Glasses Mens Womens Tortoise Bold Rectangular Full Frame Anti Glare grade ckbdp9088
Window Tint Film Chevy (back doors) DIY
Union 3" Female Ports Stainless Steel Pipe Fitting
Ebe Women Reading Glasses Reader Cheaters Anti Reflective Lenses TR90 ry2209

Recommended:
Mediven Sheer and Soft 15-20 mmHg Thigh w/ Lace Silicone Top Band CT Wheat II - Ankle 8-8.75 inches
MightySkins Skin Decal Wrap Compatible with Apple Sticker Protective Cover 100's of Color Options
MightySkins Skin Decal Wrap Compatible with DJI Sticker Protective Cover 100's of Color Options
3 1/2"W x 20"D x 20"H Funston Craftsman Smooth Bracket, Douglas Fir
MightySkins Skin Decal Wrap Compatible with HP Sticker Protective Cover 100's of Color Options
Owlpack Clear Poly Bags with Open End, 1.5 Mil, Perfect for Products, Merchandise, Goody Bags, Party Favors (4x4 inches)
Ebe Women Reading Glasses Reader Cheaters Anti Reflective Lenses TR90 ry2209
Handcrafted Ercolano Music Box Featuring "Luncheon of the Boating Party" by Renoir, Pierre Auguste - New YorkNew York
A6 Invitation Envelopes w/Peel & Press (4 3/4 x 6 1/2) - Baby Blue (1000 Qty.)
MightySkins Skin Decal Wrap Compatible with Lenovo Sticker Protective Cover 100's of Color Options

Many recommendations align with the known positives. This provides further validation. This hybrid recommendation engine can now get recommendations for all other users.

Summary

This chapter discussed hybrid recommendation engines and how they can overcome the shortfalls of other types of engines. It also showcased the implementation with the help of LightFM.

## CHAPTER 7

Clustering-Based Recommender Systems

Recommender systems based on unsupervised machine learning algorithms are very popular because they overcome many challenges that collaborative, hybrid, and classification-based systems face. A clustering technique is used to recommend the products/items based on the patterns and behaviors captured within each segment/cluster. This technique is good when data is limited, and there is no labeled data to work with.

Unsupervised learning is a machine learning category where labeled data is not leveraged, but still, inferences are discovered using the data at hand. Let's find the patterns without the dependent variables to solve business problems. Figure 7-1 shows the clustering outcome.

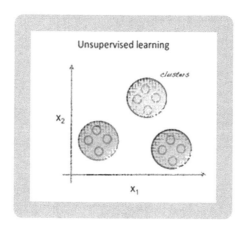

Figure 7-1. *Clustering*

Grouping similar things into segments is called *clustering*; in our terms, "things" are not data points but a collection of observations. They are

© Akshay Kulkarni, Adarsha Shivananda, Anoosh Kulkarni, V Adithya Krishnan 2023
A. Kulkarni et al., *Applied Recommender Systems with Python*, https://doi.org/10.1007/978-1-4842-8954-9_7

- Similar to each other in the same group

- Dissimilar to the observations in other groups

There are mainly two important algorithms that are highly being used in the industry. Before getting into the projects, let's briefly examine how algorithms work.

Approach

The following basic steps build a model based on similar users' recommendations.

1. Data collection

2. Data preprocessing

3. Exploratory data analysis

4. Model building

5. Recommendations

Figure 7-2 shows the step to building the clustering-based model.

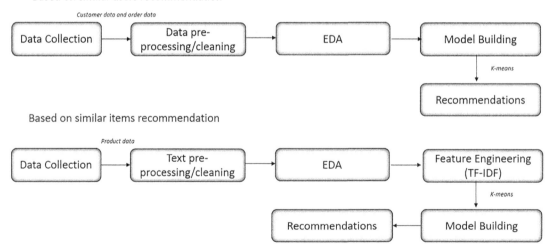

Figure 7-2. *Steps*

Implementation

Let's install and import the required libraries.

```
#Importing the libraries

import pandas as pd
import numpy as np

from matplotlib import pyplot as plt
from scipy.cluster.hierarchy import dendrogram

from sklearn.cluster import AgglomerativeClustering
from sklearn.cluster import KMeans

import seaborn as sns
import os

from sklearn import preprocessing
```

Data Collection and Downloading Required Word Embeddings

Let's consider an e-commerce dataset. Download the dataset from the GitHub link of this book.

Importing the Data as a DataFrame (pandas)

Import the records, customers, and product data.

```
# read Record dataset
df_order = pd.read_excel("Rec_sys_data.xlsx")

#read Customer Dataset
df_customer = pd.read_excel("Rec_sys_data.xlsx", sheet_name = 'customer')

# read product dataset
df_product = pd.read_excel("Rec_sys_data.xlsx", sheet_name = 'product')
```

Print the top five rows of the DataFrame.

```
#Viewing Top 5 Rows
print(df_order.head())
print(df_customer.head())
print(df_product.head())
```

Figure 7-3 shows the output of the first five rows of records data.

	InvoiceNo	StockCode	Quantity	InvoiceDate	DeliveryDate	Discount%	ShipMode	ShippingCost	CustomerID
0	536365	84029E	6	2010-12-01 08:26:00	2010-12-02 08:26:00	0.20	ExpressAir	30.12	17850
1	536365	71053	6	2010-12-01 08:26:00	2010-12-02 08:26:00	0.21	ExpressAir	30.12	17850
2	536365	21730	6	2010-12-01 08:26:00	2010-12-03 08:26:00	0.56	Regular Air	15.22	17850
3	536365	84406B	8	2010-12-01 08:26:00	2010-12-03 08:26:00	0.30	Regular Air	15.22	17850
4	536365	22752	2	2010-12-01 08:26:00	2010-12-04 08:26:00	0.57	Delivery Truck	5.81	17850

Figure 7-3. *The output*

Figure 7-4 shows the output of the first five rows of customer data.

	CustomerID	Gender	Age	Income	Zipcode	Customer Segment
0	13089	male	53	High	8625	Small Business
1	15810	female	22	Low	87797	Small Business
2	15556	female	29	High	29257	Corporate
3	13137	male	29	Medium	97818	Middle class
4	16241	male	36	Low	79200	Small Business

Figure 7-4. *The output*

Figure 7-5 shows the output of the first five rows of product data.

	StockCode	Product Name	Description	Category	Brand	Unit Price
0	22629	Ganma Superheroes Ordinary Life Case For Samsu...	New unique design, great gift.High quality pla...	Cell Phones|Cellphone Accessories|Cases & Prot...	Ganma	13.99
1	21238	Eye Buy Express Prescription Glasses Mens Wome...	Rounded rectangular cat-eye reading glasses. T...	Health|Home Health Care|Daily Living Aids	Eye Buy Express	19.22
2	22181	MightySkins Skin Decal Wrap Compatible with Ni...	Each Nintendo 2DS kit is printed with super-hi...	Video Games|Video Game Accessories|Accessories...	Mightyskins	14.99
3	84879	Mediven Sheer and Soft 15-20 mmHg Thigh w/ Lac...	The sheerest compression stocking in its class...	Health|Medicine Cabinet|Braces & Supports	Medi	62.38
4	84836	Stupell Industries Chevron Initial Wall D cor	Features: -Made in the USA. -Sawtooth hanger o...	Home Improvement|Paint|Wall Decals|All Wall De...	Stupell Industries	35.99

Figure 7-5. *The output*

Preprocessing the Data

Before building any model, the initial step is to clean and preprocess the data.

Let's analyze, clean, and merge the three datasets so that the merged DataFrame can be used to build ML models.

First, focus all customer data analysis to recommend products based on similar users.

Next, write a function and check for missing values in the customer data.

```python
# function to check missing values
def missing_zero_values_table(df):
        zero_val = (df == 0.00).astype(int).sum(axis=0)
        mis_val = df.isnull().sum()
        mis_val_percent = 100 * df.isnull().sum() / len(df)
        mz_table = pd.concat([zero_val, mis_val, mis_val_percent], axis=1)
        mz_table = mz_table.rename(
        columns = {0 : 'Zero Values', 1 : 'Missing Values', 2 : '% of Total
        Values'})
        mz_table['Total Zero Missing Values'] = mz_table['Zero Values'] +
        mz_table['Missing Values']
        mz_table['% Total Zero Missing Values'] = 100 * mz_table['Total
        Zero Missing Values'] / len(df)
        mz_table['Data Type'] = df.dtypes
        mz_table = mz_table[
            mz_table.iloc[:,1] != 0].sort_values(
        '% of Total Values', ascending=False).round(1)
        print ("Your selected dataframe has " + str(df.shape[1]) + "
        columns and " + str(df.shape[0]) + " Rows.\n"
            "There are " + str(mz_table.shape[0]) +
             " columns that have missing values.")
#        mz_table.to_excel('D:/sampledata/missing_and_zero_values.xlsx',
        freeze_panes=(1,0), index = False)
        return mz_table

# let us call the function now
missing_zero_values_table(df_customer)
```

Figure 7-6 shows the missing values output.

```
Your selected dataframe has 6 columns and 4372 Rows.
There are 0 columns that have missing values.
```

Zero Values   Missing Values   % of Total Values   Total Zero Missing Values   % Total Zero Missing Values   Data Type

Figure 7-6. *The output*

Exploratory Data Analysis

Let's explore the data for visualization using the Matplotlib package defined in sklearn.

First, let's look at age distribution.

```
# Count of age Category
plt.figure(figsize=(10,6))
plt.title("Ages Frequency")
sns.axes_style("dark")
sns.violinplot(y=df_customer["Age"])
plt.show()
```

Figure 7-7 shows the age distribution output.

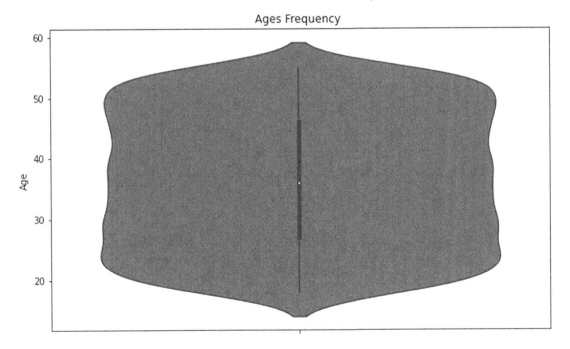

Figure 7-7. *The output*

Next, let's look at gender distribution.

```
# Count of gender Category
genders = df_customer.Gender.value_counts()
sns.set_style("darkgrid")
plt.figure(figsize=(10,4))
sns.barplot(x=genders.index, y=genders.values)
plt.show()
```

Figure 7-8 shows the gender count output.

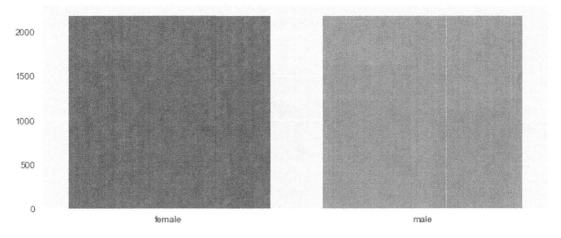

Figure 7-8. *The output*

The key insight from this chart is that data is not biased based on gender.

Let's create buckets of age columns and plot them against the number of customers.

```
# age buckets against number of customers
age18_25 = df_customer.Age[(df_customer.Age <= 25) & (df_customer.
Age >= 18)]
age26_35 = df_customer.Age[(df_customer.Age <= 35) & (df_customer.
Age >= 26)]
age36_45 = df_customer.Age[(df_customer.Age <= 45) & (df_customer.
Age >= 36)]
age46_55 = df_customer.Age[(df_customer.Age <= 55) & (df_customer.
Age >= 46)]
age55above = df_customer.Age[df_customer.Age >= 56]
```

```
x = ["18-25","26-35","36-45","46-55","55+"]
y = [len(age18_25.values),len(age26_35.values),len(age36_45.
values),len(age46_55.values),len(age55above.values)]

plt.figure(figsize=(15,6))
sns.barplot(x=x, y=y, palette="rocket")
plt.title("Number of Customer and Ages")
plt.xlabel("Age")
plt.ylabel("Number of Customer")
plt.show()
```

Figure 7-9 shows the age column buckets plotted against the number of customers.

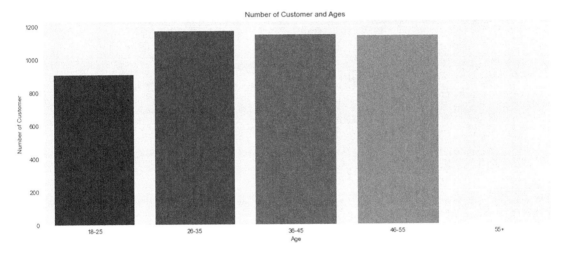

Figure 7-9. *The output*

This analysis shows that there are fewer customers ages 18 to 25.

Label Encoding

Let's encode all categorical variables.

```
# label_encoder object knows how to understand word labels.
gender_encoder = preprocessing.LabelEncoder()
segment_encoder = preprocessing.LabelEncoder()
income_encoder =  preprocessing.LabelEncoder()

# Encode labels in column
```

```
df_customer['age'] = df_customer.Age
df_customer['gender']= gender_encoder.fit_transform(df_customer['Gender'])
df_customer['customer_segment']= segment_encoder.fit_transform(df_
customer['Customer Segment'])
df_customer['income_segment']= income_encoder.fit_transform(df_
customer['Income'])

print("gender_encoder",df_customer['gender'].unique())
print("segment_encoder",df_customer['customer_segment'].unique())
print("income_encoder",df_customer['income_segment'].unique())
```

The following is the output.

```
gender_encoder [1 0]
segment_encoder [2 0 1]
income_encoder [0 1 2]
```

Let's look at the DataFrame after encoding the values.

```
df_customer.iloc[:,6:]
```

Figure 7-10 shows the output of the DataFrame after encoding the values.

	age	gender	customer_segment	income_segment
0	53	1	2	0
1	22	0	2	1
2	29	0	0	0
3	29	1	1	2
4	36	1	2	1
...	...	...	...	...
4367	22	0	0	0
4368	23	1	1	0
4369	40	1	1	2
4370	37	1	1	2
4371	19	0	1	2

4372 rows × 4 columns

Figure 7-10. *The output*

Model Building

This phase builds clusters using k-means clustering. To define an optimal number of clusters, you can also consider the elbow method or the dendrogram method.

K-Means Clustering

k-means clustering is an efficient and widely used technique that groups the data based on the distance between the points. Its objective is to minimize total variance within the cluster, as shown in Figure 7-11.

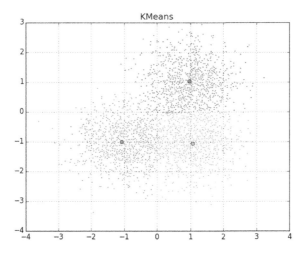

Figure 7-11. *k-means clustering*

The following steps generate clusters.

1. Use the elbow method to identify the optimum number of clusters. This acts as k.

2. Select random k points as cluster centers from the overall observations or points.

3. Calculate the distance between these centers and other points in the data and assign it to the closest center cluster that a particular point belongs to using any of the following distance metrics.

- Euclidean distance

- Manhattan distance

- Cosine distance

- Hamming distance

4. Recalculate the cluster center or centroid for each cluster.

Repeat steps 2, 3, and 4 until the same points are assigned to each cluster, and the cluster centroid is stabilized.

The Elbow Method

The elbow method checks the consistency of clusters. It finds the ideal number of clusters in data. Explained variance considers the percentage of variance explained and derives an ideal number of clusters. Suppose the deviation percentage explained is compared with the number of clusters. In that case, the first cluster adds a lot of information, but at some point, explained variance decreases, giving an angle on the graph. At the moment, the number of clusters is selected.

The elbow method runs k-means clustering on the dataset for a range of values for k (e.g., from 1–10), and then for each value of k, it computes an average score for all clusters.

Hierarchical Clustering

Hierarchical clustering is another type of clustering technique that also uses distance to create the groups. The following steps generate clusters.

1. Hierarchical clustering starts by creating each observation or point as a single cluster.

2. It identifies the two observations or points that are closest together based on the distance metrics discussed earlier.

3. Combine these two most similar points and form one cluster.

4. This continues until all the clusters are merged and form a final single cluster.

5. Finally, using a dendrogram, decide the ideal number of clusters.

The tree is cut to decide the number of clusters. The tree cutting happens where there is a maximum jump from one level to another, as shown in Figure 7-12.

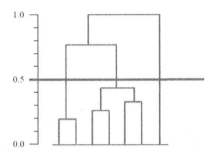

Figure 7-12. *Hierarchical clustering*

Usually, the distance between two clusters has been computed based on Euclidean distance. Many other distance metrics can be leveraged to do the same.

Let's build a *k*-means model for this use case. Before building the model, let's execute the elbow method and the dendrogram method to find the optimal clusters.

The following is an elbow method implementation.

```
# Elbow method

wcss = []
for k in range(1,15):
    kmeans = KMeans(n_clusters=k, init="k-means++")
    kmeans.fit(df_customer.iloc[:,6:])
    wcss.append(kmeans.inertia_)
plt.figure(figsize=(12,6))
plt.grid()
plt.plot(range(1,15),wcss, linewidth=2, color="red", marker ="8")
plt.xlabel("K Value")
plt.xticks(np.arange(1,15,1))
plt.ylabel("WCSS")
plt.show()print("income_encoder",df_customer['income_segment'].unique())
```

Figure 7-13 shows the elbow method output.

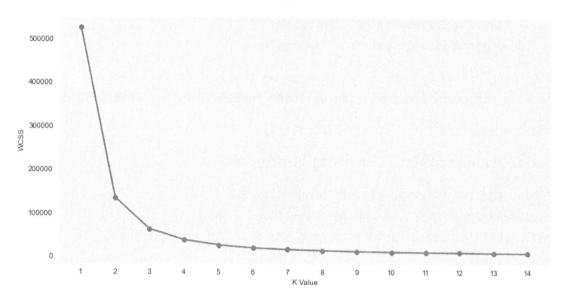

Figure 7-13. *The output*

The following is a dendrogram method implementation.

```
#function to plot dendrogram

def plot_dendrogram(model, **kwargs):
    # Create linkage matrix and then plot the dendrogram

    # create the counts of samples under each node
    counts = np.zeros(model.children_.shape[0])
    n_samples = len(model.labels_)
    for i, merge in enumerate(model.children_):
        current_count = 0
        for child_idx in merge:
            if child_idx < n_samples:
                current_count += 1  # leaf node
            else:
                current_count += counts[child_idx - n_samples]
        counts[i] = current_count

    linkage_matrix = np.column_stack(
        [model.children_, model.distances_, counts]
```

```
    ).astype(float)

    # Plot the corresponding dendrogram
    dendrogram(linkage_matrix, **kwargs)

# setting distance_threshold=0 ensures we compute the full tree.
model = AgglomerativeClustering(distance_threshold=0, n_clusters=None)

model = model.fit(df_customer.iloc[:,6:])

plt.title("Hierarchical Clustering Dendrogram")

# plot the top three levels of the dendrogram
plot_dendrogram(model, truncate_mode="level", p=3)
plt.xlabel("Number of points in node (or index of point if no
parenthesis).")
plt.show()
```

Figure 7-14 shows the dendrogram output.

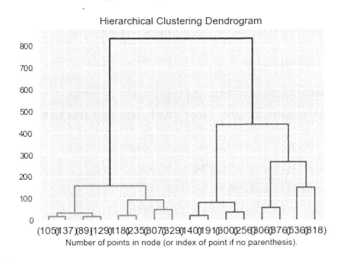

Figure 7-14. *The output*

The optimal or least number of clusters for both methods is two. But let's consider 15 clusters for this use case.

> **Note**   You can consider any number of clusters for implementation, but it should be greater than the optimal, or the least, number of clusters from *k*-means clustering or dendrogram.

Let's build a *k*-means algorithm considering 15 clusters.

```
# K-means

# Perform kmeans

km = KMeans(n_clusters=15)
clusters = km.fit_predict(df_customer.iloc[:,6:])

# saving prediction back to raw dataset
df_customer['cluster'] = clusters

df_customer
```

Figure 7-15 shows the output of df_cluster after creating the clusters.

	CustomerID	Gender	Age	Income	Zipcode	Customer Segment	age	gender	customer_segment	income_segment	cluster
0	13089	male	53	High	8625	Small Business	53	1	2	0	3
1	15810	female	22	Low	87797	Small Business	22	0	2	1	11
2	15556	female	29	High	29257	Corporate	29	0	0	0	5
3	13137	male	29	Medium	97818	Middle class	29	1	1	2	5
4	16241	male	36	Low	79200	Small Business	36	1	2	1	14
...	...	...	...	...	...	...	...	...	...	...	...
4367	17763	female	22	High	57980	Corporate	22	0	0	0	11
4368	16078	male	23	High	38622	Middle class	23	1	1	0	11
4369	13270	male	40	Medium	57985	Middle class	40	1	1	2	6
4370	15562	male	37	Medium	91274	Middle class	37	1	1	2	14
4371	13302	female	19	Medium	79580	Middle class	19	0	1	2	4

4372 rows × 11 columns

Figure 7-15. *The output*

Select the required columns from the dataset.

```
df_customer = df_customer[['CustomerID', 'Gender', 'Age', 'Income',
'Zipcode', 'Customer Segment', 'cluster']]

df_customer
```

Figure 7-16 shows the output of df_cluster after selecting particular columns.

	CustomerID	Gender	Age	Income	Zipcode	Customer Segment	cluster
0	13089	male	53	High	8625	Small Business	3
1	15810	female	22	Low	87797	Small Business	11
2	15556	female	29	High	29257	Corporate	5
3	13137	male	29	Medium	97818	Middle class	5
4	16241	male	36	Low	79200	Small Business	14
...	...	...	...	...	...	...	...
4367	17763	female	22	High	57980	Corporate	11
4368	16078	male	23	High	38622	Middle class	11
4369	13270	male	40	Medium	57985	Middle class	6
4370	15562	male	37	Medium	91274	Middle class	14
4371	13302	female	19	Medium	79580	Middle class	4

4372 rows × 7 columns

Figure 7-16. *The output*

Let's perform some analysis on the cluster level.

Write a function to plot charts of clusters against the column given.

```python
def plotting_percentages(df, col, target):
    x, y = col, target

    # Temporary dataframe with percentage values
    temp_df = df.groupby(x)[y].value_counts(normalize=True)
    temp_df = temp_df.mul(100).rename('percent').reset_index()

    # Sort the column values for plotting
    order_list = list(df[col].unique())
    order_list.sort()

    # Plot the figure
    sns.set(font_scale=1.5)
    g = sns.catplot(x=x, y='percent', hue=y,kind='bar', data=temp_df,
                height=8, aspect=2, order=order_list, legend_out=False)
    g.ax.set_ylim(0,100)

    # Loop through each bar in the graph and add the percentage value
    for p in g.ax.patches:
        txt = str(p.get_height().round(1)) + '%'
        txt_x = p.get_x()
```

```
        txt_y = p.get_height()
        g.ax.text(txt_x,txt_y,txt)

    # Set labels and title
    plt.title(f'{col.title()} By Percent {target.title()}',
            fontdict={'fontsize': 30})
    plt.xlabel(f'{col.title()}', fontdict={'fontsize': 20})
    plt.ylabel(f'{target.title()} Percentage', fontdict={'fontsize': 20})
    plt.xticks(rotation=75)
    return g
```

Plot the customer segment.

```
plotting_percentages(df_customer, 'cluster', 'Customer Segment')
```

Figure 7-17 shows the plot for the customer segment against clusters.

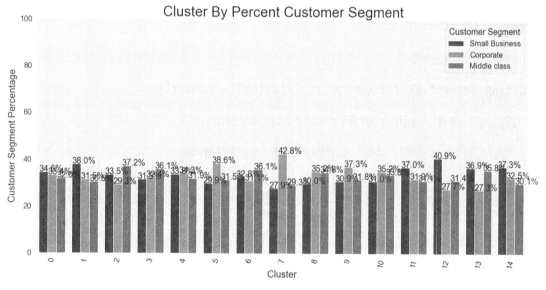

Figure 7-17. *The output*

Let's plot income.

```
plotting_percentages(df_customer, 'cluster', 'Income')
```

Figure 7-18 shows the plot for income against clusters.

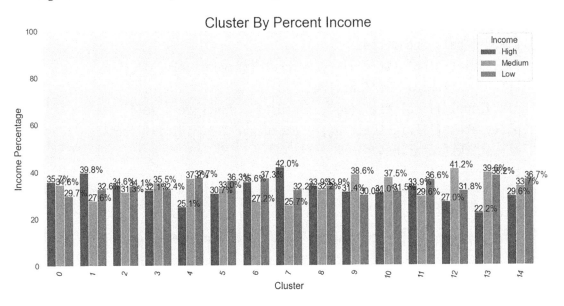

Figure 7-18. *The output*

Let's plot for gender.

```
plotting_percentages(df_customer, 'cluster', 'Gender')
```

Figure 7-19 shows the plot for gender against clusters.

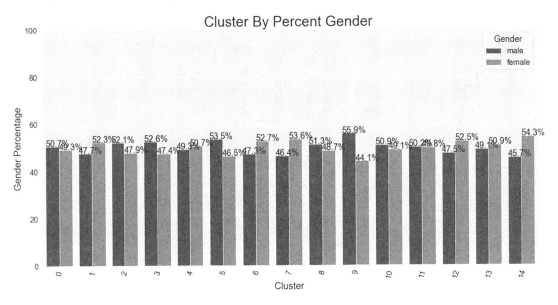

Figure 7-19. *The output*

Let's plot a chart that gives the average age per cluster.

```
df_customer.groupby('cluster').Age.mean().plot(kind='bar')
```

Figure 7-20 shows the plot for average age per cluster.

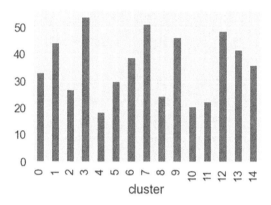

Figure 7-20. *The output*

Until now, all the data preprocessing, EDA, and model building have been performed on customer data.

Next, join customer data with the order data to get the product ID for each record.

```
order_cluster_mapping = pd.merge( df_order,df_customer, on='CustomerID',
how='inner')[['StockCode','CustomerID','cluster']]

order_cluster_mapping
```

Figure 7-21 shows the output after merging customer data with order data.

	StockCode	CustomerID	cluster
0	84029E	17850	12
1	71053	17850	12
2	21730	17850	12
3	84406B	17850	12
4	22752	17850	12
...	...	...	...
272399	21818	17666	13
272400	21817	17666	13
272401	23329	17666	13
272402	71459	17666	13
272403	84949	17666	13

272404 rows × 3 columns

Figure 7-21. *The output*

Now, let's create score_df using groupby on 'cluster' and 'StockCode', and count it.

```
score_df = order_cluster_mapping.groupby(['cluster','StockCode']).count().
reset_index()

score_df = score_df.rename(columns={'CustomerID':'Score'})

score_df
```

Figure 7-22 shows the output after creating score_df.

	cluster	StockCode	Score
0	0	10002	5
1	0	10125	3
2	0	10133	12
3	0	10135	10
4	0	11001	2
...	...	...	...
37027	14	90209C	1
37028	14	90210B	1
37029	14	M	6
37030	14	PADS	2
37031	14	POST	58

37032 rows × 3 columns

Figure 7-22. *The output*

The score_df data is ready to recommend new products to a customer. Other customers in the same cluster have bought the recommended products. This is based on similar users.

Let's focus on *product data* to recommend products based on similarity.

The preprocessing function for customer analysis is used to check the missing values.

```
missing_zero_values_table(df_product)
```

Figure 7-23 shows the missing value output.

	Zero Values	Missing Values	% of Total Values	Total Zero Missing Values	% Total Zero Missing Values	Data Type
StockCode	0	25954	86.8	25954	86.8	object
Product Name	0	25954	86.8	25954	86.8	object
Description	0	25954	86.8	25954	86.8	object
Brand	0	1129	3.8	1129	3.8	object
Category	0	792	2.6	792	2.6	object
Unit Price	0	118	0.4	118	0.4	float64

Figure 7-23. *The output*

So, there are discrepancies present in the product data. Let's clean it and check again.

```
df_product = df_product.dropna()
missing_zero_values_table(df_product)
```

169

Figure 7-24 shows the output after removing missing values.

```
Your selected dataframe has 6 columns and 3706 Rows.
There are 0 columns that have missing values.
```

Zero Values	Missing Values	% of Total Values	Total Zero Missing Values	% Total Zero Missing Values	Data Type

Figure 7-24. *The output*

Let's work on the Description column since we're dealing with similar items.

The Description column contains text, so preprocessing and converting text to features are required.

```
# Pre-processing step: remove words like we'll, you'll, they'll etc.
df_product['Description'] = df_product['Description'].replace({"'ll": " "},
regex=True)

df_product['Description'] = df_product['Description'].replace({"-": " "},
regex=True)

df_product['Description'] = df_product['Description'].replace({"[^A-Za-z0-9
]+": ""}, regex=True)


# Converting text to features
# Create word vectors from combined frames
# Make sure to make necessary imports

from sklearn.cluster import KMeans
from sklearn import metrics
from sklearn.feature_extraction.text import TfidfVectorizer

#converting text to features

vectorizer = TfidfVectorizer(stop_words='english')
X = vectorizer.fit_transform(df_product['Description'])
```

The text preprocessing and text-to-features conversion are done. Now let's build a *k*-means model using 15 clusters.

```
# #clustering your products based on text

km_des = KMeans(n_clusters=15,init='k-means++')
clusters = km_des.fit_predict(X)

df_product['cluster'] = clusters

df_product
```

Figure 7-25 shows the output after creating clusters for product data.

	StockCode	Product Name	Description	Category	Brand	Unit Price	cluster
0	22629	Ganma Superheroes Ordinary Life Case For Samsu...	New unique design great giftHigh quality plast...	Cell Phones|Cellphone Accessories|Cases & Prot...	Ganma	13.99	2
1	21238	Eye Buy Express Prescription Glasses Mens Wome...	Rounded rectangular cat eye reading glasses Th...	Health|Home Health Care|Daily Living Aids	Eye Buy Express	19.22	0
2	22181	MightySkins Skin Decal Wrap Compatible with Ni...	Each Nintendo 2DS kit is printed with super hi...	Video Games|Video Game Accessories|Accessories...	Mightyskins	14.99	6
3	84879	Mediven Sheer and Soft 15-20 mmHg Thigh w/ Lac...	The sheerest compression stocking in its class...	Health|Medicine Cabinet|Braces & Supports	Medi	62.38	9
4	84836	Stupell Industries Chevron Initial Wall D cor	Features Made in the USA Sawtooth hanger on ...	Home Improvement|Paint|Wall Decals|All Wall De...	Stupell Industries	35.99	8
...	...	...	...	...	...	...	...
3953	84612B	Home Cardboard Flower Print Travel Memo Collec...	Special design easy to insert and remove your ...	Arts, Crafts & Sewing|Scrapbooking|Albums & Re...	Unique Bargains	20.99	2
3954	47502	6 1/4 x 6 1/4 Gatefold Invitation - Mandarin O...	Announce your event using a classic Gatefold s...	Office|Envelopes & Mailing Supplies|Envelopes	Envelopes.com	55.23	2
3955	84546	Three Things That Makes Good Coffee: Sugar, Su...	Product FeaturesSize 35in x 18inColor Light pi...	Home Improvement|Paint|Wall Decals|All Wall De...	Style & Apply	39.95	3
3956	21923	Women's Breeze Walker	Supple leather uppers with lining three adjust...	Clothing|Shoes|Womens Shoes|All Womens Shoes	Prop?t	76.95	10
3957	16161M	LG PTAC 15 100 BTU/Cooling 11 900 BTU/Heating ...	LG PTAC 15 100 BTUCooling 11 900 BTUHeating He...	Home Improvement|Heating, Cooling, & Air Quali...	LG	5000.00	2

3706 rows × 7 columns

Figure 7-25. *The output*

Now the df_product data is ready to recommend the products based on similar items.

Let's write a function that recommends products based on item and user similarity.

```
# functions to recommend products based on item and user similarity.

from sklearn.feature_extraction.text import TfidfVectorizer, ENGLISH_
STOP_WORDS
from sklearn.metrics.pairwise import cosine_similarity
from sklearn.feature_extraction.text import TfidfTransformer
```

```python
from nltk.corpus import stopwords
import pandas as pd

# function to find cosine similarity after converting discerption column to
features using TF-IDF
def cosine_similarity_T(df,query):

    vec = TfidfVectorizer(analyzer='word', stop_words=ENGLISH_STOP_WORDS)
    vec_train = vec.fit_transform(df.Description)
    vec_query = vec.transform([query])

    within_cosine_similarity = []

    for i in range(len(vec_train.todense())):
        within_cosine_similarity.append(cosine_similarity(vec_train[i,:].
        toarray(), vec_query.toarray())[0][0])

    df['Similarity'] = within_cosine_similarity

    return df

def recommend_product(customer_id):
# filter for the particular customer
    cluster_score_df = score_df[score_df.cluster==order_cluster_
    mapping[order_cluster_mapping.CustomerID == customer_id]['cluster'].
    iloc[0]]

# filter top 5 stock codes for recommendation
    top_5_non_bought = cluster_score_df[~cluster_score_df.StockCode.
    isin(order_cluster_mapping[order_cluster_mapping.CustomerID ==
    customer_id]['StockCode'])].nlargest(5, 'Score')
    print('\n--- top 5 StockCode - Non bought --------\n')
    print(top_5_non_bought)

    print('\n-------Recommendations Non bought ------\n')

#printing product names from product table.   print(df_product[df_product.
StockCode.isin(top_5_non_bought.StockCode)]['Product Name'])
```

```python
    cust_orders = df_order[df_order.CustomerID == customer_id]
    [['CustomerID','StockCode']]

    top_orders = cust_orders.groupby(['StockCode']).count().reset_index()
    top_orders = top_orders.rename(columns = {'CustomerID':'Counts'})
    top_orders['CustomerID'] = customer_id

    top_5_bought = top_orders.nlargest(5,'Counts')

    print('\n--- top 5 StockCode - bought --------\n')

    print(top_5_bought)

    print('\n-------Stock code Product (Bought) - Description cluster
    Mapping------\n')
    top_clusters = df_product[df_product.StockCode.isin(top_5_bought.
    StockCode.tolist())][['StockCode','cluster']]
    print(top_clusters)

    df = df_product[df_product['cluster']==df_product[df_product.
    StockCode==top_clusters.StockCode.iloc[0]]['cluster'].iloc[0]]
    query = df_product[df_product.StockCode==top_clusters.StockCode.
    iloc[0]]['Description'].iloc[0]

    print("\nquery\n")

    print(query)

    recomendation = cosine_similarity_T(df,query)

    print(recomendation.nlargest(3,'Similarity'))

recommend_product(13137)
```

Figure 7-26 highlights the final recommendations to customer 13137.

```
--- top 5 StockCode - Non bought --------

        cluster StockCode  Score
15032         5    85123A    122
14490         5     47566    115
13533         5     22423     95
12686         5     21034     62
13246         5     22077     59
```

```
-------Recommendations Non bought ------

135      Mediven Sheer and Soft 15-20 mmHg Thigh w/ Lac...
215      MightySkins Skin Decal Wrap Compatible with Ap...
225      Handcrafted Ercolano Music Box Featuring "Lunc...
741      3 Pack Newbee Fashion- "Butterfly" Thin Design...
1048     Port Authority K110 Dry Zone UV Micro-Mesh Pol...
Name: Product Name, dtype: object
```

```
--- top 5 StockCode - bought --------

     StockCode  Counts  CustomerID
23      21212        5       13137
24      21213        5       13137
86      22211        5       13137
101     22379        5       13137
8       20727        4       13137
```

```
-------Stock code Product (Bought) - Description cluster Mapping------

      StockCode  cluster
214      21212        4
372      22379        2
565      20727       14
636      22211        2
1129     21213        8
```

```
     StockCode                                    Description  Similarity
44       84378  Our Rustic Collection is an instant classic Ou...         1.0
111      23298  Our Rustic Collection is an instant classic Ou...         1.0
214      21212  Our Rustic Collection is an instant classic Ou...         1.0
```

Figure 7-26. *The output*

The first set highlights similar user recommendations. The second set highlights similar item recommendations.

Summary

In this chapter, you learned how to build a recommendation engine using unsupervised ML algorithms, which is clustering. Customer and order data were used to recommend the products/items based on similar users. The product data was used to recommend the products/items using similar items.

CHAPTER 8

Classification Algorithm–Based Recommender Systems

A classification algorithm-based recommender system is also known as the *buying propensity model*. The goal here is to predict the propensity of customers to buy a product using historical behavior and purchases.

The more accurately you predict future purchases, the better recommendations and, in turn, sales. This kind of recommender system is used more often to ensure 100% conversion from the users who are likely to purchase with certain probabilities. Promotions are offered on those products, enticing users to make a purchase.

Approach

The following basic steps build a classification algorithm-based recommender engine.

1. Data collection

2. Data preprocessing and cleaning

3. Feature engineering

4. Exploratory data analysis

5. Model building

6. Evaluation

7. Predictions and recommendations

Figure 8-1 shows the steps for building a classification algorithm-based model.

© Akshay Kulkarni, Adarsha Shivananda, Anoosh Kulkarni, V Adithya Krishnan 2023
A. Kulkarni et al., *Applied Recommender Systems with Python*, https://doi.org/10.1007/978-1-4842-8954-9_8

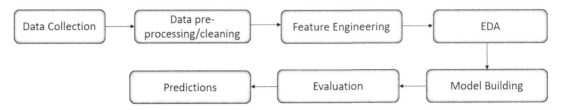

Figure 8-1. *Classification-based model*

Implementation

Let's install and import the required libraries.

```
#Importing the libraries

import pandas as pd
import numpy as np
import matplotlib.pyplot as plt
import seaborn as sns.display import Image
import os
from sklearn import preprocessing
from sklearn.model_selection import train_test_split
from sklearn.metrics import accuracy_score,confusion_
matrix,classification_report
from sklearn.linear_model import LogisticRegression
from imblearn.combine import SMOTETomek
from collections import Counter
from sklearn.ensemble import RandomForestClassifier
from sklearn.metrics import roc_auc_score, roc_curve
from sklearn.neighbors import KNeighborsClassifier
from sklearn.metrics import roc_curve, roc_auc_score
from sklearn.naive_bayes import GaussianNB
from sklearn import tree
from sklearn.tree import DecisionTreeClassifier
from xgboost import XGBClassifier
```

Data Collection and Download Word Embeddings

Let's consider an e-commerce dataset. Download the dataset from GitHub link.

Importing the Data as a DataFrame (pandas)

Import the records, customers, and product data.

```
# read Record dataset
record_df = pd.read_excel("Rec_sys_data.xlsx")

#read Customer Dataset
customer_df = pd.read_excel("Rec_sys_data.xlsx", sheet_name = 'customer')

# read product dataset
prod_df = pd.read_excel("Rec_sys_data.xlsx", sheet_name = 'product')
```

Print the top five rows of the DataFrame.

```
#Viewing Top 5 Rows
print(record_df.head())
print(customer_df.head())
print(prod_df.head())
```

Figure 8-2 shows the output of the first five rows of records data.

	InvoiceNo	StockCode	Quantity	InvoiceDate	DeliveryDate	Discount%	ShipMode	ShippingCost	CustomerID
0	536365	84029E	6	2010-12-01 08:26:00	2010-12-02 08:26:00	0.20	ExpressAir	30.12	17850
1	536365	71053	6	2010-12-01 08:26:00	2010-12-02 08:26:00	0.21	ExpressAir	30.12	17850
2	536365	21730	6	2010-12-01 08:26:00	2010-12-03 08:26:00	0.56	Regular Air	15.22	17850
3	536365	84406B	8	2010-12-01 08:26:00	2010-12-03 08:26:00	0.30	Regular Air	15.22	17850
4	536365	22752	2	2010-12-01 08:26:00	2010-12-04 08:26:00	0.57	Delivery Truck	5.81	17850

Figure 8-2. *The output*

Figure 8-3 shows the output of the first five rows of customer data.

	CustomerID	Gender	Age	Income	Zipcode	Customer Segment
0	13089	male	53	High	8625	Small Business
1	15810	female	22	Low	87797	Small Business
2	15556	female	29	High	29257	Corporate
3	13137	male	29	Medium	97818	Middle class
4	16241	male	36	Low	79200	Small Business

Figure 8-3. *The output*

Figure 8-4 shows the output of the first five rows of product data.

	StockCode	Product Name	Description	Category	Brand	Unit Price
0	22629	Ganma Superheroes Ordinary Life Case For Samsu...	New unique design, great gift.High quality pla...	Cell Phones|Cellphone Accessories|Cases & Prot...	Ganma	13.99
1	21238	Eye Buy Express Prescription Glasses Mens Wome...	Rounded rectangular cat-eye reading glasses. T...	Health|Home Health Care|Daily Living Aids	Eye Buy Express	19.22
2	22181	MightySkins Skin Decal Wrap Compatible with Ni...	Each Nintendo 2DS kit is printed with super-hi...	Video Games|Video Game Accessories|Accessories...	Mightyskins	14.99
3	84879	Mediven Sheer and Soft 15-20 mmHg Thigh w/ Lac...	The sheerest compression stocking in its class...	Health|Medicine Cabinet|Braces & Supports	Medi	62.38
4	84836	Stupell Industries Chevron Initial Wall D cor	Features: -Made in the USA. -Sawtooth hanger o...	Home Improvement|Paint|Wall Decals|All Wall De...	Stupell Industries	35.99

Figure 8-4. *The output*

Preprocessing the Data

Before building any model, the initial step is to clean and preprocess the data.

Analyze, clean, and merge the three datasets so that the merged DataFrame can build ML models.

Now, let's check what the total quantity of each product taken by each customer is.

```
# group By Stockcode and CustomerID and sum the Quantity
group = pd.DataFrame(record_df.groupby(['StockCode', 'CustomerID']).
Quantity.sum())
print(group.shape)
group.head()
```

Figure 8-5 shows the output of grouping by stock code and customer ID and sums the quantity.

```
(192758, 1)
```

		Quantity
StockCode	CustomerID	
10002	12451	12
	12510	24
	12583	48
	12637	12
	12673	1

Figure 8-5. *The output*

Next, check for null values for customers and records datasets.

```
#Check for null values
print(record_df.isnull().sum())
print("-------------\n")
print(customer_df.isnull().sum())
```

The following is the output.

```
InvoiceNo      0
StockCode      0
Quantity       0
InvoiceDate    0
DeliveryDate   0
Discount%      0
ShipMode       0
ShippingCost   0
CustomerID     0
dtype: int64
-------------

CustomerID     0
Gender         0
Age            0
Income         0
Zipcode        0
```

```
Customer Segment    0
dtype: int64
```

There are no null values present in the datasets. So, dropping or treating them is not required.

Let's load CustomerID and StockCode into different variables and create a cross-product for further usage.

```
#Loading the CustomerID and StockCode into different variable d1, d2
d2 = customer_df['CustomerID']
d1 = record_df["StockCode"]

# Taking the sample of data and storing into two variables
row = d1.sample(n= 900)
row1 = d2.sample(n=900)

# Cross product of row and row1
index = pd.MultiIndex.from_product([row, row1])
a = pd.DataFrame(index = index).reset_index()
a.head()
```

Figure 8-6 shows the output.

	StockCode	CustomerID
0	48129	13736
1	48129	17252
2	48129	16005
3	48129	17288
4	48129	14267

Figure 8-6. *The output*

Now, let's merge 'group' and 'a' with 'CustomerID' and 'StockCode'.

```
#merge customerID and StockCode
data = pd.merge(group,a, on = ['CustomerID', 'StockCode'], how = 'right')
data.head()
```

Figure 8-7 shows the output.

	StockCode	CustomerID	Quantity
0	48129	13736	NaN
1	48129	17252	NaN
2	48129	16005	1.0
3	48129	17288	NaN
4	48129	14267	NaN

Figure 8-7. *The output*

As you can see, null values are present in the Quantity column.

Let's check for nulls.

```
#check total number of null values in quantity column
print(data['Quantity'].isnull().sum())
# check the shape of data that is number of rows and columns
print(data.shape)
```

The following is the output.

```
779771
(810000, 3)
```

Let's treat missing values by replacing null with zeros and checking for unique values.

```
#replacing nan values with 0
data['Quantity'] = data['Quantity'].replace(np.nan, 0).astype(int)
```

```
# Check all unique value of quantity column
print(data['Quantity'].unique())
```

Figure 8-8 shows the output.

```
[      0      12     24      2      1     72      3      5      4     48     10      8
       6      32     14     96     37      7     20     30      9     15     11     58
      41      18    400     40    100    220     16     64    120    144     84     13
      36     192     42     25     38     50    168    240     52     19     21    101
      35      60    161    648    480    576     45    600    156     17     46     44
     232     385    180     22     43   1200    200    348    148     76     63     28
     360      54    276     31    370     98    122    706     26    456    960     23
     160     104    401     57    135     27    179    136    224     80    376     97
     201     145     90     70    154   1000     75    432    132    152   2160    624
     720     390     88     34    108    125     33    250    225    133    408     95
     384     130    528    896    514    174    300     62     85    110    150    264
      49     184     68    126   1062    288    549     78    301    112    289    312
     241     732   1400    252    320     56    109    275     29    190    102     66
     660     185    216    119    504     74    256     61    266    128     83   4300
      71      79    664    500     55     73    270    208    768    257     53    138
     172     114    242     47    121    230    111    163    273     59     51   1152
     302      67    984    350    512    832    322   1248   1800   1680   1296    123
     140     147    198   1540    481    336    672    792    204    210    800    448
    9360      94    285    483    450    151    105    750    286   1920    372    680
     324     175    577   1008    212    431     65    640    124    170    194     93
      81    2400    196    234    588    386    304   1080    840     91    248    193
   74215      39    864]
```

Figure 8-8. *The output*

Let's now drop unnecessary columns from the product table.

```
## drop product name and description column
product_data = prod_df.drop(['Product Name', 'Description'], axis = 1)
product_data['Category'].str.split('::').str[0]
product_data.head()
```

Figure 8-9 shows the output of the first five rows.

	StockCode	Category	Brand	Unit Price
0	22629	Cell Phones|Cellphone Accessories|Cases & Prot...	Ganma	13.99
1	21238	Health|Home Health Care|Daily Living Aids	Eye Buy Express	19.22
2	22181	Video Games|Video Game Accessories|Accessories...	Mightyskins	14.99
3	84879	Health|Medicine Cabinet|Braces & Supports	Medi	62.38
4	84836	Home Improvement|Paint|Wall Decals|All Wall De...	Stupell Industries	35.99

Figure 8-9. *The output*

Let's extract the first hierarchy level from the category column and join the product_ data table.

```
# extract the first string category column
cate = product_data['Category'].str.extract(r"(\w+)", expand=True)

# join cat column with original dataset
df2 = product_data.join(cate, lsuffix="_left")
df2.drop(['Category'], axis = 1, inplace = True)

# rename column to Category
df2 = df2.rename(columns = {0: 'Category'})
print(df2.shape)
df2.head()
```

Figure 8-10 shows the output.

```
(29912, 4)
```

	StockCode	Brand	Unit Price	Category
0	22629	Ganma	13.99	Cell
1	21238	Eye Buy Express	19.22	Health
2	22181	Mightyskins	14.99	Video
3	84879	Medi	62.38	Health
4	84836	Stupell Industries	35.99	Home

Figure 8-10. *The output*

Let's check and drop null values, if any, after joining.

```
#check for null values and drop it
df2.isnull().sum()
df2.dropna(inplace = True)
df2.isnull().sum()
```

The following is the output.

```
StockCode     0
Brand         0
Unit Price    0
Category      0
```

```
dtype: int64
```

Save the preprocessed file and read it again.

```
## save to csv file
df2.to_csv("Products.csv")
```

```
# Load product dataset
product = pd.read_csv("/content/Products.csv")
```

Merge the data, product, and customer tables.

```
## Merge data and product dataset
final_data = pd.merge(data, product, on= 'StockCode')
```

```
# create final dataset by merging customer & final data
final_data1 = pd.merge(customer_df, final_data, on = 'CustomerID')
```

```
# Drop Unnamed and zipcode column
final_data1.drop(['Unnamed: 0', 'Zipcode'], axis = 1, inplace = True)
final_data1.head()
```

Figure 8-11 shows the output of the first five rows after merging.

	CustomerID	Gender	Age	Income	Customer Segment	StockCode	Quantity	Brand	Unit Price	Category
0	16241	male	36	Low	Small Business	84997A	0	Mightyskins	23.99	Electronics
1	16241	male	36	Low	Small Business	M	0	Dr. Comfort	139.00	Clothing
2	16241	male	36	Low	Small Business	85032B	1	Mightyskins	59.99	Sports
3	16241	male	36	Low	Small Business	85170B	0	Mediven	62.38	Health
4	16241	male	36	Low	Small Business	85099F	0	Tom Ford	63.20	Beauty

Figure 8-11. *The output*

Check for null values in the final table.

```
print(final_data1.shape)
# Check for null values in each columns
final_data1.isnull().sum()
```

The following is the output.

```
(61200, 10)
```

```
CustomerID          0
Gender              0
Age                 0
Income              0
Customer Segment    0
StockCode           0
Quantity            0
Brand               0
Unit Price          0
Category            0
dtype: int64
```

Check the unique categories in each column.

```
#Check for unique value in each categorical columns

print(final_data1['Category'].unique())
print('------------\n')
print(final_data1['Income'].unique())
print('------------\n')
print(final_data1['Brand'].unique())
print('------------\n')
print(final_data1['Customer Segment'].unique())
print('------------\n')
print(final_data1['Gender'].unique())
print('------------\n')
print(final_data1['Quantity'].unique())
```

The following is the output.

```
['Electronics' 'Clothing' 'Sports' 'Health' 'Beauty' 'Jewelry' 'Home'
 'Office' 'Auto' 'Cell' 'Pets' 'Food' 'Household' 'Shop']
------------

['Low' 'Medium' 'High']
------------

['Mightyskins' 'Dr. Comfort' 'Mediven' 'Tom Ford' 'Eye Buy Express'
```

```
 'MusicBoxAttic' 'Duda Energy' 'Business Essentials' 'Medi'
 'Seat Belt Extender Pros' 'Boss (hub)' 'Ishow Hair' 'Ekena Milwork'
 'JustVH' 'UNOTUX' 'Envelopes.com' 'Auburn Leathercrafters'
 'Style & Apply' 'Edwards' 'Larissa Veronica' 'Awkward Styles' 'New Way'
 'McDonalds' 'Ekena Millwork' 'Omega' "Medaglia D'Oro" 'allwitty' 'Prop?t'
 'Unique Bargains' 'CafePress' "Ron's Optical" 'Wrangler' 'AARCO']
------------

['Small Business' 'Middle class' 'Corporate']
------------

['male' 'female']
------------

[    0    1    3    5   15    2    4    8    6   24    7   30    9   10
    62   20   18   12   72   50  400   36   27  242   58   25   60   48
    22  148   16  152   11   31   64  147   42   23   43   26   14   21
  1200  500   28  112   90  128   44  200   34   96  140   19  160   17
   100  320  370  300  350   32   78  101   66   29]
```

From this output, you can see some special characters in the brand column. Let's remove them.

```
## test cleaning
final_data1['Brand'] = final_data1['Brand'].str.replace('?', '')
final_data1['Brand'] = final_data1['Brand'].str.replace('&', 'and')
final_data1['Brand'] = final_data1['Brand'].str.replace('(', '')
final_data1['Brand'] = final_data1['Brand'].str.replace(')', '')
print(final_data1['Brand'].unique())
```

The following is the output.

```
['Mightyskins' 'Dr. Comfort' 'Mediven' 'Tom Ford' 'Eye Buy Express'
 'MusicBoxAttic' 'Duda Energy' 'Business Essentials' 'Medi'
 'Seat Belt Extender Pros' 'Boss hub' 'Ishow Hair' 'Ekena Milwork'
 'JustVH' 'UNOTUX' 'Envelopes.com' 'Auburn Leathercrafters'
 'Style and Apply' 'Edwards' 'Larissa Veronica' 'Awkward Styles' 'New Way'
 'McDonalds' 'Ekena Millwork' 'Omega' "Medaglia D'Oro" 'allwitty' 'Propt'
 'Unique Bargains' 'CafePress' "Ron's Optical" 'Wrangler' 'AARCO']
```

All the datasets have merged, and the required data preprocessing and cleaning are completed.

Feature Engineering

Once the data is preprocessed and cleaned, the next step is to perform feature engineering.

Let's create a flag column, using the Quantity column, that indicates whether the customer has bought the product or not.

If the Quantity column is 0, the customer has not bought the product.

```
#creating buy_falg column
final_data1.loc[final_data1.Quantity == 0 ,"flag_buy" ] = 0
final_data1.loc[final_data1.Quantity != 0 ,"flag_buy" ] = 1

# Converting the values of flag_buy column into integer
final_data1['flag_buy'] = final_data1.flag_buy.astype(int)
final_data1.tail()
```

Figure 8-12 shows the first five rows' output after creating the target column.

	CustomerID	Gender	Age	Income	Customer Segment	StockCode	Quantity	Brand	Unit Price	Category	flag_buy
61195	16078	male	23	High	Middle class	15044D	0	Awkward Styles	18.95	Clothing	0
61196	16078	male	23	High	Middle class	90082B	0	Ron's Optical	11.99	Health	0
61197	16078	male	23	High	Middle class	72802C	0	Wrangler	44.99	Clothing	0
61198	16078	male	23	High	Middle class	82494L	0	AARCO	512.92	Office	0
61199	16078	male	23	High	Middle class	84341B	0	Medi	62.38	Health	0

Figure 8-12. *The output*

A new flag_buy column is created. Let's do some basic exploration of that column.

```
#Check for the unique value in flag buy column
print(final_data1['flag_buy'].unique())
# Gives the description of columns
print(final_data1.describe())
##Information about the data
print(final_data1.info())
```

Figure 8-13 shows the description output.

```
array([0, 1])
```

	CustomerID	Age	Quantity	Unit Price	flag_buy
count	61200.000000	61200.000000	61200.000000	61200.000000	61200.000000
mean	15246.732222	36.741111	0.584673	75.782941	0.033840
std	1715.931502	10.855742	9.865284	85.066391	0.180818
min	12346.000000	18.000000	0.000000	4.150000	0.000000
25%	13742.000000	27.000000	0.000000	25.740000	0.000000
50%	15260.500000	37.000000	0.000000	49.580000	0.000000
75%	16686.250000	46.000000	0.000000	64.990000	0.000000
max	18280.000000	55.000000	1200.000000	512.920000	1.000000

Figure 8-13. *The output*

```
<class 'pandas.core.frame.DataFrame'>
Int64Index: 61200 entries, 0 to 61199
Data columns (total 11 columns):
 #   Column            Non-Null Count   Dtype
---  ------            --------------   -----
 0   CustomerID        61200 non-null   int64
 1   Gender            61200 non-null   object
 2   Age               61200 non-null   int64
 3   Income            61200 non-null   object
 4   Customer Segment  61200 non-null   object
 5   StockCode         61200 non-null   object
 6   Quantity          61200 non-null   int64
 7   Brand             61200 non-null   object
 8   Unit Price        61200 non-null   float64
 9   Category          61200 non-null   object
 10  flag_buy          61200 non-null   int64
dtypes: float64(1), int64(4), object(6)
memory usage: 5.6+ MB
```

Exploratory Data Analysis

Feature engineering is a must for model data preprocessing. However, exploratory data analysis (EDA) also plays a vital role.

You can get more business insights by looking at the historical data itself.

Let's start exploring the data. Plot a chart of the brand column.

```
plt.figure(figsize=(50,20))
sns.set_theme(style="darkgrid")
sns.countplot(x = 'Brand', data = final_data1)
```

Figure 8-14 shows the output of the brand column.

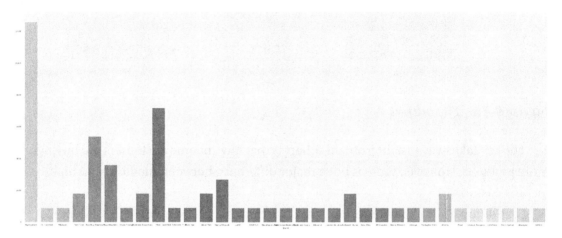

Figure 8-14. *The output*

The key insight from this chart is that the Mightyskins brand has the highest sales.

Let's plot the Income column.

```
# Count of Income Category
plt.figure(figsize=(10,5))
sns.set_theme(style="darkgrid")
sns.countplot(x = 'Income', data = final_data1)
```

Figure 8-15 shows the count chart's Income column output.

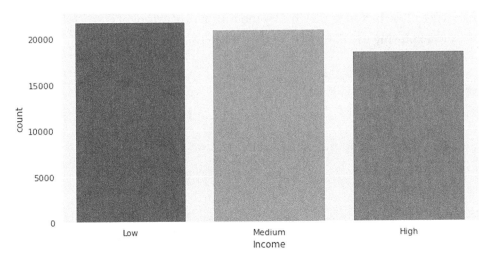

Figure 8-15. *The output*

The key takeaway insight from this chart is that low-income customers are buying more products. However, there is not a major difference between medium and high-income customers.

Let's dump a few charts here. For more information, please refer to the notebook.

Plot a histogram to show age distribution.

```
# histogram plot to show distribution age
plt.figure(figsize=(10,5))
sns.set_theme(style="darkgrid")
sns.histplot(data=final_data1, x="Age", kde = True)
```

Figure 8-16 shows the age distribution output.

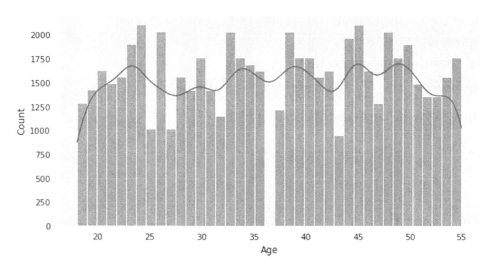

Figure 8-16. *The output*

Plot an area chart to show age distribution with hue by category.

```
plt.figure(figsize=(10,5))
sns.set_theme(style="darkgrid")
sns.histplot(data=final_data1, x="Age", hue="Category", element= "poly")
```

Figure 8-17 shows the age distributions by category.

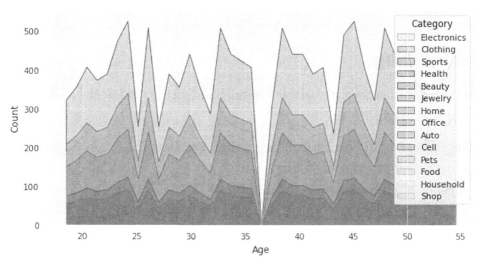

Figure 8-17. *The output*

Create a bar plot to check the target distribution.

```
# Count plot to show number of customer bought the product
plt.figure(figsize=(10,5))
sns.set_theme(style="darkgrid")
sns.countplot(x = 'flag_buy', data = final_data1)
```

Figure 8-18 is the target distribution bar plot.

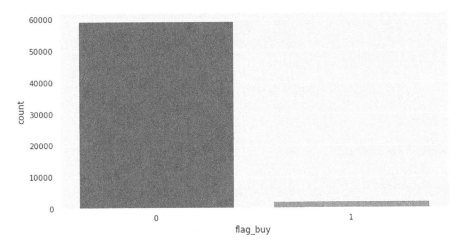

Figure 8-18. *The output*

It looks like this particular use case has a data imbalance. Let's build the model after sampling the data.

Model Building

Let's encode all the categorical variables before building the model. Also, store the stock code for further usage.

```
#Encoding categorical variable using Label Encoder
from sklearn import preprocessing
label_encoder = preprocessing.LabelEncoder()
final_data1['StockCode'] = label_encoder.fit_transform(final_
data1['StockCode'])
mappings = {}
mappings['StockCode'] = dict(zip(label_encoder.classes_,range(len(label_
encoder.classes_))))
final_data1['Gender'] = label_encoder.fit_transform(final_data1['Gender'])
```

```
final_data1['Customer Segment'] = label_encoder.fit_transform(final_
data1['Customer Segment'])
final_data1['Brand'] = label_encoder.fit_transform(final_data1['Brand'])
final_data1['Category'] = label_encoder.fit_transform(final_
data1['Category'])
final_data1['Income'] = label_encoder.fit_transform(final_data1['Income'])
final_data1.head()
```

Figure 8-19 shows the first five rows after encoding.

	CustomerID	Gender	Age	Income	Customer Segment	StockCode	Quantity	Brand	Unit Price	Category	flag_buy
0	16241	1	36	1	2	24	0	20	23.99	4	0
1	16241	1	36	1	2	52	0	6	139.00	3	0
2	16241	1	36	1	2	26	1	20	59.99	13	1
3	16241	1	36	1	2	41	0	19	62.38	6	0
4	16241	1	36	1	2	36	0	28	63.20	1	0

Figure 8-19. *The output*

Train-Test Split

The data is split into two parts: one for training the model, which is the training set, and another for evaluating the model, which is the test set. The train_test_split library from sklearn.model_selection is imported to split the DataFrame into two parts.

```
## separating dependent and independent variables
x = final_data1.drop(['flag_buy'], axis = 1)
y = final_data1['flag_buy']
```

```
# check the shape of dependent and independent variable
print((x.shape, y.shape))
```

```
# splitting data into train and test
from sklearn.model_selection import train_test_split
x_train, x_test, y_train, y_test = train_test_split(x, y, train_size = 0.6,
random_state = 42)
```

Logistic Regression

Linear regression is needed to predict a numerical value. But you also encounter classification problems where dependent variables are binary, like yes or no, 1 or 0, true or false, and so on. In that case, *logistic regression* is needed. It is a classification algorithm and continues linear regression. Here, log odds are used to restrict the dependent variable between 0 and 1.

Figure 8-20 shows the logistic regression formula.

a.k.a. **Log Odds**   **Intercept**

or **Logit**

$$\log\left(\frac{P}{1-P}\right) = \beta_0 + \beta_1 X$$

Figure 8-20. Formula

Where $(P/1 - P)$ is the odds ratio, β_0 is constant, and β is the coefficient. Figure 8-21 shows how logistic regression works.

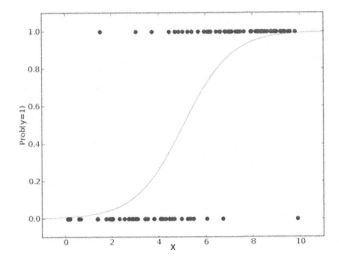

Figure 8-21. Logistic regression

Now let's look at how to evaluate the classification model.

- Accuracy is the number of correct predictions divided by the total number of predictions. The values lie between 0 and 1; to convert it into a percentage, multiply the answer by 100. But only considering accuracy as the evaluation parameter is not an ideal thing to do. For example, if the data is imbalanced, you can obtain very high accuracy.

- The crosstab between an actual and a predicted class is called a *confusion matrix*. It's not only for binary, but you can also use it for multiclass classification. Figure 8-22 represents a confusion matrix.

Predicted

		Positive	Negative
Actual	**Positive**	True Positive	False Negative
	Negative	False Positive	True Negative

***Figure 8-22.** Confusion matrix*

- The ROC (receiver operating characteristic) curve is an evaluation metric for classification tasks. A plot with a false positive rate on the x axis and a true positive rate on the y axis is the ROC curve plot. It says how strongly the classes are distinguished when the thresholds are varied. Higher the value of the area under the ROC curve, the higher the predictive power. Figure 8-23 shows the ROC curve.

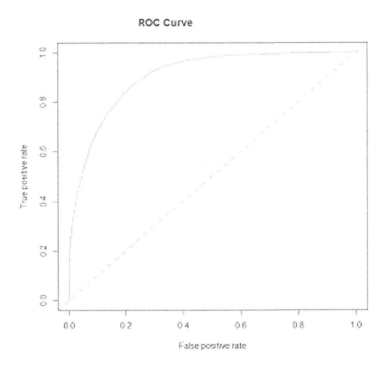

Figure 8-23. *ROC curve*

Linear and logistic regression are the traditional way of using statistics as a base to predict the dependent variable. But there are a few drawbacks to these algorithms.

- Statistical modeling must satisfy the assumptions that are discussed previously. If they are not satisfied, models won't be reliable and thoroughly fit random predictions.

- These algorithm face challenge when data and target feature is non-linear. Complex patterns are hard to decode.

- Data should be clean (missing values and outliers should be treated).

Advanced machine learning concepts like decision tree, random forest, SVM, and neural networks can be used to overcome these limitations.

Implementation

```
##training using logistic regression
```

```
from sklearn.metrics import accuracy_score,confusion_
matrix,classification_report
from sklearn.linear_model import LogisticRegression
logistic = LogisticRegression()
logistic.fit(x_train, y_train)

# calculate score
pred=logistic.predict(x_test)
print(confusion_matrix(y_test, pred))
print(accuracy_score(y_test, pred))
print(classification_report(y_test, pred))
```

The following is the output.

```
[[23633     0]
 [    2   845]]
0.9999183006535948
```

	precision	recall	f1-score	support
0	1.00	1.00	1.00	23633
1	1.00	1.00	1.00	847
accuracy			1.00	24480
macro avg	1.00	1.00	1.00	24480
weighted avg	1.00	1.00	1.00	24480

This chapter's "Exploratory Data Analysis" section discussed the target distribution and its imbalances. Let's apply a sampling technique, make it balanced data, and then build the model.

```
# Sampling technique to handle imbalanced data
smk = SMOTETomek(0.50)
X_res,y_res=smk.fit_resample(x_train,y_train)

# Count the number of classes
from collections import Counter
print("The number of classes before fit {}".format(Counter(y)))
print("The number of classes after fit {}".format(Counter(y_res)))
```

The following is the output.

```
The number of classes before fit Counter({0: 59129, 1: 2071})
The number of classes after fit Counter({0: 35428, 1: 17680})
```

Build the same model after sampling.

```
## Training model with Logistics Regression
from sklearn.metrics import accuracy_score,confusion_
matrix,classification_report
from sklearn.linear_model import LogisticRegression
logistic = LogisticRegression()
logistic.fit(X_res, y_res)

# Calculate Score
y_pred=logistic.predict(x_test)
print(confusion_matrix(y_test,y_pred))
print(accuracy_score(y_test,y_pred))
print(classification_report(y_test,y_pred))
```

The following is the output.

```
[[23633     0]
 [    0   847]]

1.0
```

	precision	recall	f1-score	support
0	1.00	1.00	1.00	23633
1	1.00	1.00	1.00	847
accuracy			1.00	24480
macro avg	1.00	1.00	1.00	24480
weighted avg	1.00	1.00	1.00	24480

Decision Tree

The decision is a type of supervised learning in which the data is split into similar groups based on the most important variable to the least. It looks like a tree-shaped structure when all the variables split hence the name tree-based models.

The tree comprises a root node, a decision node, and a leaf node. A decision node can have two or more branches, and a leaf node represents a decision. Decision trees handle any type of data, be it quantitative or qualitative. Figure 8-24 shows how the decision tree works.

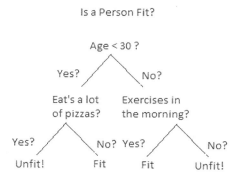

Figure 8-24. *Decision tree*

Let's examine how tree splitting happens, which is the key concept in decision trees. The core of the decision tree algorithm is the process of splitting the tree. It uses different algorithms to split the node and is different for classification and regression problems.

The following are for classification problems.

- The *Gini index* is a probabilistic way of splitting the trees. It uses the sum of the probability square for success and failure and decides the purity of the nodes. CART (classification and regression tree) uses the Gini index to create splits.

- *Chi-square* is the statistical significance between subnodes, and the parent node decides the splitting. Chi-square = ((actual – expected)^2 / expected)^1/2. CHAID (Chi-square Automatic Interaction Detector) is an example of this.

The following pertains to regression problems.

- *Reduction in variance* works based on the variance between two features (target and independent feature) to split a tree.

- *Overfitting* occurs when the algorithms tightly fit the given training data but is inaccurate in predicting the outcomes of the untrained or test data. The same is the case with decision trees as well. It occurs when the tree is created to perfectly fit all samples in the training dataset, affecting test data accuracy.

Implementation

```
##Training model using decision tree
from sklearn import tree
from sklearn.tree import DecisionTreeClassifier
dt = DecisionTreeClassifier()
dt.fit(X_res, y_res)
y_pred = dt.predict(x_test)
print(dt.score(x_train, y_train))
print(confusion_matrix(y_test,y_pred))
print(accuracy_score(y_test,y_pred))
print(classification_report(y_test,y_pred))
```

The following is the output.

```
1.0


[[23633     0]
 [    0   847]]

1.0
```

	precision	recall	f1-score	support
0	1.00	1.00	1.00	23633
1	1.00	1.00	1.00	847
accuracy			1.00	24480
macro avg	1.00	1.00	1.00	24480
weighted avg	1.00	1.00	1.00	24480

Random Forest

Random forest is the most widely used machine learning algorithm because of its flexibility and ability to overcome the overfitting problem. A random forest is an ensemble algorithm that is an ensemble of multiple decision trees. The higher the number of trees, the better the accuracy.

The random forest can perform both classification and regression tasks. The following are some of its advantages.

- It is insensitive to missing values and outliers.

- It prevents the algorithm from overfitting.

How does it work? It works on bagging and bootstrap sample techniques.

- Randomly takes the square root of m features and 2/3 bootstrap data sample with a replacement for training each decision tree randomly and predicts the outcome

- Builds n number of trees until the out-of-bag error rate is minimized and stabilized

- Computes the votes for each predicted target and considers the mode as a final prediction in terms of classification

Figure 8-25 shows the working of the random forest model.

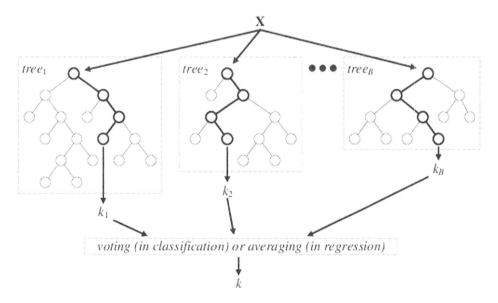

Figure 8-25. *Random forest*

Implementation

```
##Training model using Random forest
from sklearn.ensemble import RandomForestClassifier
rf = RandomForestClassifier()
rf.fit(X_res, y_res)

# Calculate Score
y_pred=rf.predict(x_test)
print(confusion_matrix(y_test,y_pred))
print(accuracy_score(y_test,y_pred))
print(classification_report(y_test,y_pred))
```

The following is the output.

```
[[23633     0]
 [    0   847]]
1.0
              precision    recall  f1-score   support

           0       1.00      1.00      1.00     23633
           1       1.00      1.00      1.00       847

    accuracy                           1.00     24480
   macro avg       1.00      1.00      1.00     24480
weighted avg       1.00      1.00      1.00     24480
```

KNN

For more information on the algorithm, please refer to Chapter 4.

Implementation

```
#Training model using KNN
from sklearn.metrics import roc_auc_score, roc_curve
from sklearn.neighbors import KNeighborsClassifier
model1 = KNeighborsClassifier(n_neighbors=3)
model1.fit(X_res,y_res)
y_predict = model1.predict(x_test)
```

```
# Calculate Score
print(model1.score(x_train, y_train))
print(confusion_matrix(y_test,y_predict))
print(accuracy_score(y_test,y_predict))
print(classification_report(y_test,y_predict))

# plot AUROC curve
r_auc = roc_auc_score(y_test, y_predict)
r_fpr, r_tpr, _ = roc_curve(y_test, y_predict)
plt.plot(r_fpr, r_tpr, linestyle='--', label='KNN prediction
(AUROC = %0.3f)' % r_auc)
plt.title('ROC Plot')
# Axis labels
plt.xlabel('False Positive Rate')
plt.ylabel('True Positive Rate')
# Show legend
plt.legend()
# Show plot
plt.show()
```

Figure 8-26 shows the KNN output.

```
0.9875272331154684
[[22929   704]
 [  225   622]]
0.9620506535947713
                    precision    recall  f1-score   support

               0       0.99      0.97      0.98     23633
               1       0.47      0.73      0.57       847

        accuracy                           0.96     24480
       macro avg .     0.73      0.85      0.78     24480
    weighted avg       0.97      0.96      0.97     24480
```

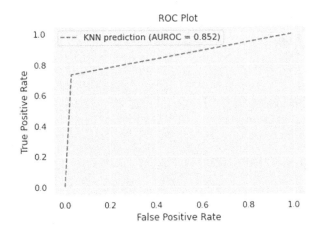

Figure 8-26. *The output*

Note   Naive Bayes and XGBoost implementations are also in the notebooks.

In the preceding models, the logistic regression performance is better than all other models.

So, using that model, let's recommend the products to one customer.

```
# x_test has all the features, lets us take the copy of it
test_data = x_test.copy()

#let us store predictions in one variable
test_data['predictions'] = pred

#filter the data and recommend.
```

```
recomm_one_cust = test_data[(test_data['CustomerID']== 17315) & (test_
data['predictions']== 1)]
```

```
# to build the model we have encoded the stockcode column now we will
decode and recommend.
items = []
for item_id in recomm_one_cust['StockCode'].unique().tolist():
    prod = {v: k for k, v in mappings['StockCode'].items()}[item_id]
    items.append(str(prod))
```

```
items
```

The following is the output.

```
['85123A', '85099C', '84970L', 'POST', '84970S', '82494L', '48173C',
'85099B']
```

These are the product IDs that should be recommended for customer 17315.

If you want recommendations with product names, filter these IDs in the product table.

```
recommendations = []
for i in items:
    recommendations.append(prod_df[prod_df['StockCode']== i]
['Product Name'])
```

```
recommendations
```

The following is the output.

```
[135     Mediven Sheer and Soft 15-20 mmHg Thigh w/ Lac...
 Name: Product Name, dtype: object,
 551     Mediven Sheer and Soft 15-20 mmHg Thigh w/ Lac...
 Name: Product Name, dtype: object,
 1282     Eye Buy Express Kids Childrens Reading Glasses...
 Name: Product Name, dtype: object,
 7    MightySkins Skin Decal Wrap Compatible with Ot...
 Name: Product Name, dtype: object,
 160     Union 3" Female Ports Stainless Steel Pipe Fit...
 Name: Product Name, dtype: object,
```

```
179     AARCO Enclosed Wall Mounted Bulletin Board
Name: Product Name, dtype: object,
287      Mediven Sheer and Soft 15-20 mmHg Thigh w/ Lac...
Name: Product Name, dtype: object,
77      Ebe Women Reading Glasses Reader Cheaters Anti...
Name: Product Name, dtype: object]
```

You can also do this recommendation using the probability output from the model by sorting them.

Summary

In this chapter, you learned how to recommend a product/item to the customers using various classification algorithms, from data cleaning to model building. These types of recommendations are an add-on to the e-commerce platform. With classification-based algorithm output, you can show the hidden products to the user, and the customer is more likely to be interested in those products/items. The conversion rate of these recommendations is high compared to other recommender techniques.

## CHAPTER 9

Deep Learning–Based Recommender System

So far, you have learned various methods for building recommender systems and saw their implementation in Python. The book began with basic and intuitive methods, like market basket analysis, arithmetic-based content, and collaborative filtering methods, and then moved on to more complex machine learning methods, like clustering, matrix factorizations, and machine learning classification-based methods. This chapter continues the journey by implementing an end-to-end recommendation system using advanced deep learning concepts.

Deep learning techniques utilize recent and rapidly growing network architectures and optimization algorithms to train on large amounts of data and build more expressive and better-performing models. Graphics Processing Units (GPUs) and deep learning have been driving advances in recommender systems for the past few years. Due to their massively parallel architecture, using GPUs for computation provides higher performance and cost savings. Let's first explore the basics of deep learning and then look at the deep learning–based collaborative filtering method (neural collaborative filtering).

Basics of Deep Learning (ANNs)

Deep learning is a subtype of machine learning, which essentially covers algorithms based upon artificial neural networks (ANN), a network with connecting nodes resembling the neural links present in a biological brain. It is a set of connected nodes (artificial neurons) that transmit information through links or edges. It starts with an input layer (of nodes) and branches out into multiple layers of nodes, called the hidden layers, before rejoining into a single output node/layer, which gets the output predictions.

© Akshay Kulkarni, Adarsha Shivananda, Anoosh Kulkarni, V Adithya Krishnan 2023
A. Kulkarni et al., *Applied Recommender Systems with Python*, https://doi.org/10.1007/978-1-4842-8954-9_9

Figure 9-1 shows a neural network, the building blocks of any deep learning algorithm.

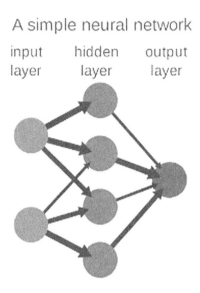

Figure 9-1. *Neural network*

Each node's output transforms the input based on the weights provided to nodes and edges, and as you progress through the network, from input to output layer, the prediction is optimized and refined further by each layer. This is known as *forward propagation*. Another important process, known as *backpropagation*, uses loss optimization algorithms, such as gradient descent, to calculate and reduce the losses in prediction by adjusting the weights of the nodes and edges for each layer while moving backward from the output layer toward the input layer. These two processes work together to build the final network that gives accurate predictions.

This was a simple explanation of basic neural networks, which are typically the building blocks of every deep learning algorithm.

Neural Collaborative Filtering (NCF)

Collaborative filtering methods have been the most popular for building recommendation systems in various domains. Popular techniques like matrix factorization have been extensively used because they are easy to implement and provide accurate predictions. But in recent times, through new areas of research, deep learning–based models are being used increasingly in all domains, including collaborative filtering.

Neural collaborative filtering (NCF) is a supercharged advanced collaborative filtering method using neural networks. In matrix factorization, the user-item relationship is defined through the inner product of user and item matrices. In NCF, this inner product is replaced by a neural network structure. Through this, it tries to achieve two things.

- Generalize the matrix factorization into a neural network framework

- Learn the user-item interactions/relationship through a multilayer perceptron (MLP)

Figure 9-2 shows the overall NCF structure.

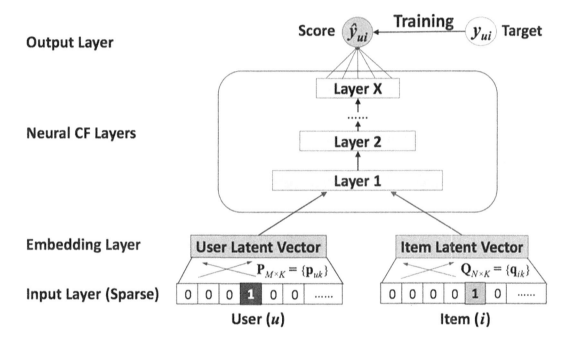

Figure 9-2. *NCF*

Multilayer perceptron (MLP) is a neural network with multiple layers that are fully connected (i.e., all nodes of the previous layer are connected to all nodes in the next layer). Every node in an MLP usually uses a sigmoid function as its activation function. The sigmoid function takes real values as input and returns a real value between 0 and 1 using this formula: sigmoid(x) = 1/(1 + exp(–x)), where x is the input. In NCF, the activation function is a rectified linear activation function (ReLU). The input returns the same number if it's a positive value; it outputs a 0 if it's a negative value. The formula for ReLU is max(0, x), where x is the input.

Figure 9-3 shows a multilayer perceptron (MLP).

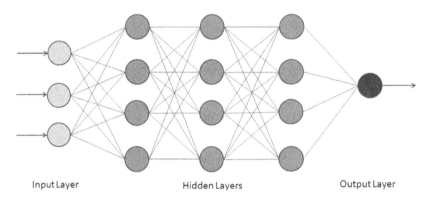

Figure 9-3. *Multilayer perceptron (MLP)*

The use of an MLP algorithm over a matrix factorization algorithm is surely an upgrade, as theoretically, an MLP can learn any continuous relationship with much greater accuracy, and it has a high level of non-linearity (due to multilayers), making it a better fit to learn the intricate interactions between users and items.

Now that you've heard about the basics of deep learning, neural nets, and neural collaborative filtering, let's dive into an end-to-end implementation of a deep learning/NCF-based recommendation system in the following sections.

Implementation

The following installs and imports the required libraries.

```
#Importing the libraries

%load_ext autoreload
%autoreload 2

import sys
import pandas as pd
import tensorflow as tf
```

```
tf.get_logger().setLevel('ERROR') # only show error messages

from recommenders.utils.timer import Timer
from recommenders.models.ncf.ncf_singlenode import NCF
from recommenders.models.ncf.dataset import Dataset as NCFDataset
#from recommenders.datasets import movielens
from recommenders.utils.notebook_utils import is_jupyter
from recommenders.datasets.python_splitters import python_chrono_
split,python_stratified_split
from recommenders.evaluation.python_evaluation import (rmse, mae, rsquared,
exp_var, map_at_k, ndcg_at_k, precision_at_k,

                                              recall_at_k, get_
                                              top_k_items)

print("System version: {}".format(sys.version))
print("Pandas version: {}".format(pd.__version__))
print("Tensorflow version: {}".format(tf.__version__))
```

Data Collection

Let's consider an e-commerce dataset. Download the dataset from the GitHub link.

Importing the Data as a DataFrame (pandas)

Let's import the records, customers, and product data.

```
# read Record dataset
record_df = pd.read_excel("Rec_sys_data.xlsx")

#read Customer Dataset
customer_df = pd.read_excel("Rec_sys_data.xlsx", sheet_name = 'customer')

# read product dataset
prod_df = pd.read_excel("Rec_sys_data.xlsx", sheet_name = 'product')
```

Next, print the top five rows of the DataFrame.

```
#Viewing Top 5 Rows
print(record_df.head())
```

```
print(customer_df.head())
print(prod_df.head())
```

Figure 9-4 shows the output of the first five rows of records data.

	InvoiceNo	StockCode	Quantity	InvoiceDate	DeliveryDate	Discount%	ShipMode	ShippingCost	CustomerID
0	536365	84029E	6	2010-12-01 08:26:00	2010-12-02 08:26:00	0.20	ExpressAir	30.12	17850
1	536365	71053	6	2010-12-01 08:26:00	2010-12-02 08:26:00	0.21	ExpressAir	30.12	17850
2	536365	21730	6	2010-12-01 08:26:00	2010-12-03 08:26:00	0.56	Regular Air	15.22	17850
3	536365	84406B	8	2010-12-01 08:26:00	2010-12-03 08:26:00	0.30	Regular Air	15.22	17850
4	536365	22752	2	2010-12-01 08:26:00	2010-12-04 08:26:00	0.57	Delivery Truck	5.81	17850

Figure 9-4. *The output*

Figure 9-5 shows the output of the first five rows of customer data.

	CustomerID	Gender	Age	Income	Zipcode	Customer Segment
0	13089	male	53	High	8625	Small Business
1	15810	female	22	Low	87797	Small Business
2	15556	female	29	High	29257	Corporate
3	13137	male	29	Medium	97818	Middle class
4	16241	male	36	Low	79200	Small Business

Figure 9-5. *The output*

Figure 9-6 shows the output of the first five rows of product data.

	StockCode	Product Name	Description	Category	Brand	Unit Price
0	22629	Ganma Superheroes Ordinary Life Case For Samsu...	New unique design, great gift.High quality pla...	Cell Phones|Cellphone Accessories|Cases & Prot...	Ganma	13.99
1	21238	Eye Buy Express Prescription Glasses Mens Wome...	Rounded rectangular cat-eye reading glasses. T...	Health|Home Health Care|Daily Living Aids	Eye Buy Express	19.22
2	22181	MightySkins Skin Decal Wrap Compatible with Ni...	Each Nintendo 2DS kit is printed with super-hi...	Video Games|Video Game Accessories|Accessories...	Mightyskins	14.99
3	84879	Mediven Sheer and Soft 15-20 mmHg Thigh w/ Lac...	The sheerest compression stocking in its class...	Health|Medicine Cabinet|Braces & Supports	Medi	62.38
4	84836	Stupell Industries Chevron Initial Wall D cor	Features: -Made in the USA. - Sawtooth hanger o...	Home Improvement|Paint|Wall Decals|All Wall De...	Stupell Industries	35.99

Figure 9-6. *The output*

Data Preprocessing

Let's select the required columns from records_df and drop nulls, if any. Also, dropping the string item IDs (StockCode) for getting desired input data fa or modeling.

```
#selecting columns
df = record_df[['CustomerID','StockCode','Quantity','DeliveryDate']]

#dropping the StockCodes (item ids) that are string for this experiment, as
NCF only takes integer ids
df["StockCode"] = df["StockCode"].apply(lambda x: pd.to_numeric(x,
errors='coerce')).dropna()

# dropping nulls
df = df.dropna()
print(df.shape)
df
```

Figure 9-7 shows the output order data after selecting the required columns.

```
(272404, 4)
```

	CustomerID	StockCode	Quantity	DeliveryDate
0	17850	84029E	6	2010-12-02 08:26:00
1	17850	71053	6	2010-12-02 08:26:00
2	17850	21730	6	2010-12-03 08:26:00
3	17850	84406B	8	2010-12-03 08:26:00
4	17850	22752	2	2010-12-04 08:26:00
...	...	...	...	...
272399	15249	23399	12	2011-10-08 11:37:00
272400	15249	22727	4	2011-10-08 11:37:00
272401	15249	23434	12	2011-10-08 11:37:00
272402	15249	23340	12	2011-10-07 11:37:00
272403	15249	23209	10	2011-10-08 11:37:00

272404 rows × 4 columns

Figure 9-7. *The output*

Let's rename a few of the column names.

```
#header=["userID", "itemID", "rating", "timestamp"]

df = df.rename(columns={
```

```
    'CustomerID':"userID",'StockCode':"itemID",'Quantity':"rating",'Deliver
    yDate':"timestamp"
```

```
})
```

Next, change the user_id and item_id datatypes to an integer since that is the required format for NCF.

```
df["userID"] = df["userID"].astype(int)
df["itemID"] = df["itemID"].astype(int)
```

Train-Test Split

The data is split into two parts: one for training the model, which is the training set, and another for evaluating the model, which is the test set.

Let's split the data using the Spark chronological splitter provided in the utilities.

```
train, test = python_chrono_split(df, 0.75)
```

Save the train and test data into two separate files, which are later loaded into the model init function.

```
train_file = "./train.csv"
test_file = "./test.csv"
train.to_csv(train_file, index=False)
test.to_csv(test_file, index=False)
```

Modeling and Recommendations

Train the NCF model on the training data, and get the top k recommendations for our testing data. NCF accepts implicit feedback and generates a propensity of items to be recommended to users on a scale of 0 to 1. A recommended item list can then be generated based on the scores. Note that this quick-start notebook uses a smaller number of epochs to reduce the time for training. As a consequence, the model performance has deteriorated slightly.

Also, before building the model, let's define a few model parameters.

```
# top k items to recommend
```

```
TOP_K = 10

# Model parameters
EPOCHS = 50
BATCH_SIZE = 256

SEED = 42

#preparing the data
 data = NCFDataset(train_file=train_file, test_file=test_file, seed=SEED)
# training NCF model
model = NCF (
    n_users=data.n_users,
    n_items=data.n_items,
    model_type="NeuMF",
    n_factors=4,
    layer_sizes=[16,8,4],
    n_epochs=EPOCHS,
    batch_size=BATCH_SIZE,
    learning_rate=1e-3,
    verbose=10,
    seed=SEED
)

#adding timer for training.
with Timer() as train_time:
    model.fit(data)

print("Took {} seconds for training.".format(train_time))

#adding timimg for predictions
with Timer() as test_time:
    users, items, preds = [], [], []
    item = list(train.itemID.unique())
    for user in train.userID.unique():
        user = [user] * len(item)
        users.extend(user)
```

```
        items.extend(item)
        preds.extend(list(model.predict(user, item, is_list=True)))

    all_predictions = pd.DataFrame(data={"userID": users, "itemID":items,
    "prediction":preds})

    merged = pd.merge(train, all_predictions, on=["userID", "itemID"],
    how="outer")
    all_predictions = merged[merged.rating.isnull()].drop('rating', axis=1)
print("Took {} seconds for prediction.".format(test_time))
```

The following is the output.

```
Took 842.6078 seconds for training.
```

```
Took 24.8943 seconds for prediction.
```

Here, all predictions are stored in the all_predictions object.

Let's evaluate the NCF performance by using different metrics.

```
# Evaluate model
eval_map = map_at_k(test, all_predictions, col_prediction='prediction',
k=TOP_K)
eval_ndcg = ndcg_at_k(test, all_predictions, col_prediction='prediction',
k=TOP_K)
eval_precision = precision_at_k(test, all_predictions, col_
prediction='prediction', k=TOP_K)
eval_recall = recall_at_k(test, all_predictions, col_
prediction='prediction', k=TOP_K)

print("MAP:\t%f" % eval_map,
      "NDCG:\t%f" % eval_ndcg,
      "Precision@K:\t%f" % eval_precision,
      "Recall@K:\t%f" % eval_recall, sep='\n')
```

The following is the output.

```
MAP:        0.020692
NDCG:       0.064364
```

Precision@K: 0.047777
Recall@K: 0.051526

Let's read the data for the recommendation.

```
# read data
df_order = pd.read_excel('Rec_sys_data.xlsx', 'order')
df_customer = pd.read_excel('Rec_sys_data.xlsx', 'customer')
df_product = pd.read_excel('Rec_sys_data.xlsx', 'product')
```

The all_predictions object with a set of recommendations given by the model has been created.

Select required and rename the columns.

```
#select columns
all_predictions = all_predictions[['userID','itemID','prediction']]

# rename columns

all_predictions = all_predictions.rename(columns={
    "userID":'CustomerID',"itemID":'StockCode',"rating":'Quantity','prediction'
:'probability'

})
```

Now let's write a function to recommend the products by giving the customer ID as input.

The function uses the all_predictions object to recommend the products.

```
def recommend_product(customer_id):

  print(" \n---------- Top 5 Bought StockCodes -----------\n")

  print(df_order[df_order['CustomerID']==customer_id][['CustomerID',
'StockCode','Quantity']].nlargest(5,'Quantity'))

  top_5_bought = df_order[df_order['CustomerID']--customer_id][['CustomerID
','StockCode','Quantity']].nlargest(5,'Quantity')
```

```
print('\n-------Product Name of bought StockCodes ------\n')

print(df_product[df_product.StockCode.isin(top_5_bought.StockCode)]
['Product Name'])

print("\n --------- Top 5 Recommendations ------------ \n")

print(all_predictions[all_predictions['CustomerID']==customer_id].
nlargest(5,'probability'))

recommend = all_predictions[all_predictions['CustomerID']==customer_id].
nlargest(5,'probability')

print('\n-------Product Name of Recommendations ------\n')

print(df_product[df_product.StockCode.isin(recommend.StockCode)]
['Product Name'])
```

This function gets the following information.

- The top five bought stock codes (item IDs) with the product names for a given customer

- The top five recommendations for the same customer from NCF

Let's use the function to recommend products for customers 13137 and 15127.

```
recommend_product(13137)
```

Figure 9-8 shows the recommendation for customer 13137.

```
---------- Top 5 Bought StockCodes -----------

        CustomerID StockCode  Quantity
234414       13137     84077        48
234443       13137     23321        13
50797        13137     21985        12
234404       13137     22296        12
234418       13137     22297        12

-------Product Name of bought StockCodes ------

70      MightySkins Skin Decal Wrap Compatible with Li...
490            Window Tint Film Mitsubishi (all doors) DIY
694      Harriton Men's Paradise Short-Sleeve Performan...
1065     MightySkins Skin For Samsung Galaxy J3 (2016),...
1339     MightySkins Skin Decal Wrap Compatible with Le...
Name: Product Name, dtype: object

 --------- Top 5 Recommendations ------------

        CustomerID StockCode  probability
1951608      13137    85123A     0.975194
1952595      13137     21034     0.971388
1951667      13137     22197     0.960145
1951758      13137    85099F     0.929778
1952914      13137     22766     0.917395

-------Product Name of Recommendations ------

63      Tom Ford Lip Color Sheer 0.07Oz/2g New In Box ...
135     Mediven Sheer and Soft 15-20 mmHg Thigh w/ Lac...
161     Union 3" Female Ports Stainless Steel Pipe Fit...
215     MightySkins Skin Decal Wrap Compatible with Ap...
236     MightySkins Skin Decal Wrap Compatible with HP...
Name: Product Name, dtype: object
```

Figure 9-8. *The output*

```
recommend_product(15127)
```

Figure 9-9 shows the recommendations for customer 15127.

```
---------- Top 5 Bought StockCodes -----------

        CustomerID StockCode  Quantity
272296       15127     23263        48
272287       15127     23354        24
272288       15127     22813        24
272289       15127     23096        24
272285       15127     21181        12

-------Product Name of bought StockCodes ------

13                  billyboards Porcelain School Chalkboard
374         MightySkins Protective Vinyl Skin Decal for Po...
923         Zoan Synchrony Duo Sport Electric Snow Helmet ...
952         MightySkins Skin Decal Wrap Compatible with Sm...
1576        EMPIRE KLIX Klutch Designer Wallet Case for LG G2
Name: Product Name, dtype: object

--------- Top 5 Recommendations ------------

        CustomerID StockCode  probability
6135734      15127     84879     0.973742
6137006      15127     35970     0.935546
6136832      15127     21034     0.931347
6137564      15127     23356     0.925915
6137220      15127    85049A     0.922400

-------Product Name of Recommendations ------

3           Mediven Sheer and Soft 15-20 mmHg Thigh w/ Lac...
215         MightySkins Skin Decal Wrap Compatible with Ap...
288         MightySkins Skin Decal Wrap Compatible with Sm...
1558        Ebe Men Black Rectangle Half Rim Spring Hinge ...
1713        Ebe Prescription Glasses Mens Womens Burgundy ...
Name: Product Name, dtype: object
```

Figure 9-9. *The output*

Summary

This chapter covered deep learning and how deep learning–based recommendation engines work. You saw this by implementing an end-to-end deep learning–based recommender system using NF. Deep learning–based recommender systems are a very new but relevant field, and it has shown quite promising results in recent times. If sufficient data and computational access are provided, then deep learning–based techniques will surely outperform any other techniques out there in the market, and hence is a very important concept to have in your repertoire.

CHAPTER 10

Graph-Based Recommender Systems

The previous chapter covered deep learning-based recommender systems and explained how to implement end-to-end neural collaborative filtering. This chapter explores another recent advanced method: graph-based recommendation systems powered by knowledge graphs.

Figure 10-1 illustrates a graph-based recommendation system for movie recommendations.

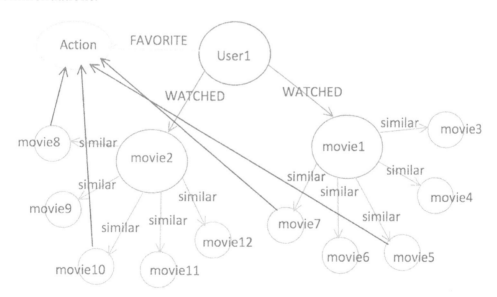

Figure 10-1. *Graph-based movie recommendations*

In graph-based recommendation systems, knowledge graph structures represent relationships between users and items. A knowledge graph is the structure of a network of interconnected datasets, enriched with semantics, that is visualized graphically

A. Kulkarni et al., *Applied Recommender Systems with Python*, https://doi.org/10.1007/978-1-4842-8954-9_10

by illustrating the relationship between multiple entities. The graph structure, when visualized, has three primary components; nodes, edges, and labels. The edge of the link defines the relationship between two nodes/entities, where each node can be any object, user, item, place, and so forth. The underlying semantics provide an additional dynamic context to the defined relationships, enabling more complex decision-making.

Figure 10-2 shows a simple one-to-one relationship in a knowledge graph structure.

A represents the subject, B represents the predicate, C
represents the object

Figure 10-2. *Simple knowledge graph connection*

Figure 10-3 explains knowledge graphs.

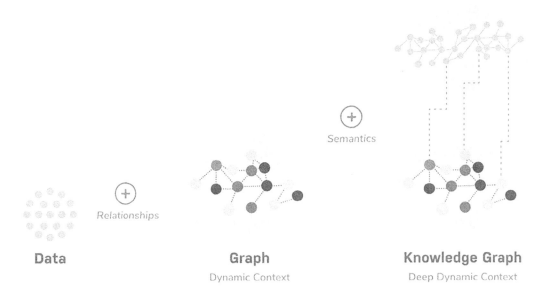

Figure 10-3. *Knowledge graphs explained*

This chapter uses Neo4j to implement the knowledge graphs. Neo4j is one of the best graph databases in the market today. It is a high-performance graph store with a user-friendly query language that is highly scalable and robust.

The knowledge graphs will fetch similar users for the required recommendations.

Implementation

The following installs and imports the required libraries.

```
# Installing required packages

!pip install py2neo
!pip install openpyxl –upgrade
!pip install neo4j
!pip install neo4jupyter

#Importing the required libraries

import pandas as pd
from neo4j import GraphDatabase, basic_auth
from py2neo import Graph
import re
import neo4jupyter
```

Before establishing the connection between Neo4j and the notebook, create a new sandbox in Neo4j at https://neo4j.com/sandbox/.

Once the sandbox is created, you must change the URL and the password.

You can find them in the connection details, as shown in Figure 10-4.

Figure 10-4. *Connection details*

Let's establish a connection between Neo4j and the Python notebook.

```
# establishing the connection

g = Graph("bolt://44.192.55.13:7687", password = "butter-ohms-chairman")

# The url "bolt://34.201.241.51:7687" needs to be replaced in case of new
sandbox creation in neo4j.
# The credentials "neo4j" and "whirls-bullet-boils" also need a replacement
for each use case.

driver = GraphDatabase.driver(
  "bolt://44.192.55.13:7687",
  auth=basic_auth("neo4j", "butter-ohms-chairman"))

def execute_transactions(transaction_execution_commands):
    # Establishing connection with database
    data_base_connection = GraphDatabase.driver
    ("bolt://44.192.55.13:7687",
    auth=basic_auth("neo4j", "butter-ohms-chairman"))
    # Creating a session
    session = data_base_connection.session()
```

```
    for i in transaction_execution_commands:
        session.run(i)
```

Let's import the data.

```
# This dataset consists of transactions which will be used to establish
relationship between the customer and the stock.
df = pd.read_excel(r'Rec_sys_data.xlsx')
```

```
# Little bit of preprocessing so that we can easily run NoSQL queries.
df['CustomerID'] = df['CustomerID'].apply(str)
```

```
# This dataset contains detailed information about each stock which will be
used to link stockcodes and their description/title.
df1 = pd.read_excel('Rec_sys_data.xlsx','product')
```

```
df1.head()
```

Figure 10-5 shows the df1 output (first five rows).

	StockCode	Product Name	Description	Category	Brand	Unit Price
0	22629	Ganma Superheroes Ordinary Life Case For Samsu...	New unique design, great gift.High quality pla...	Cell Phones|Cellphone Accessories|Cases & Prot...	Ganma	13.99
1	21238	Eye Buy Express Prescription Glasses Mens Wome...	Rounded rectangular cat-eye reading glasses. T...	Health|Home Health Care|Daily Living Aids	Eye Buy Express	19.22
2	22181	MightySkins Skin Decal Wrap Compatible with Ni...	Each Nintendo 2DS kit is printed with super-hi...	Video Games|Video Game Accessories|Accessories...	Mightyskins	14.99
3	84879	Mediven Sheer and Soft 15-20 mmHg Thigh w/ Lac...	The sheerest compression stocking in its class...	Health|Medicine Cabinet|Braces & Supports	Medi	62.38
4	84836	Stupell Industries Chevron Initial Wall D cor	Features: -Made in the USA. -Sawtooth hanger o...	Home Improvement|Paint|Wall Decals|All Wall De...	Stupell Industries	35.99

Figure 10-5. *The output*

Let's upload the entities into the Neo4j database.

To implement knowledge graphs in Neo4J, the DataFrame must be converted into a relational database. First, customers and stocks must be converted into entities (or nodes of a graph) to build a relationship between them.

```
#creating a list of all unique customer IDs
customerids = df['CustomerID'].unique().tolist()
```

```
# storing all the create commands to be executed into create_customers list
create_customers = []
```

```
for i in customerids:
```

```
  # example of create statement "create (n:entity {property_key :
  '12345'})"
    statement = "create (c:customer{cid:"+ '"' + str(i) + '"' +"})"
    create_customers.append(statement)

# running all the queries into neo4j to create customer entities
execute_transactions(create_customers)
```

Once the customer nodes are done, create nodes for the stock.

```
#  creating a list of all unique stockcodes
stockcodes = df['StockCode'].unique().tolist()

# storing all the create commands to be executed into the create_
stockcodes list
create_stockcodes = []

for i in stockcodes:
  # example of create statement "create (m:entity {property_key : 'XYZ'})"
    statement = "create (s:stock{stockcode:"+ '"' + str(i) + '"' +"})"
    create_stockcodes.append(statement)

# running all the queries into neo4j to create stock entities
execute_transactions(create_stockcodes)
```

Next, create a link between the stock codes and title, which are needed to recommend items.

For this, let's create another property key called 'Title' into the existing stock entity in our Neo4j database.

```
#creating a blank dataframe
df2 = pd.DataFrame(columns = ['StockCode', 'Title'])

#Converting stockcodes to string in both the dataframe
df['StockCode'] = df['StockCode'].astype(str)
df1['StockCode'] = df1['StockCode'].astype(str)

# This cell of code will add all the unique stockcodes along with their
title in df2
stockcodes = df['StockCode'].unique().tolist()
```

```
for i in range(len(stockcodes)):
    dict_temp = {}
    dict_temp['StockCode'] = stockcodes[i]
    dict_temp['Title'] = df1[df1['StockCode']==stockcodes[i]]['Product
    Name'].values
    temp_Df = pd.DataFrame([dict_temp])
    df2 = df2.append(temp_Df)
df2= df2.reset_index(drop=True)
# Doing some data preprocessing such that these queries can be run in neo4j
df2['Title'] = df2['Title'].apply(str)
df2['Title'] = df2['Title'].map(lambda x: re.sub(r'\W+', ' ', x))
df2['Title'] = df2['Title'].apply(str)
```

```
# This query will add the 'title' property key to each stock entity in our
neo4j database
for i in range(len(df2)):
  query = """
  MATCH (s:stock {stockcode:""" + '"' + str(df2['StockCode'][i]) +
  '"' + """})
  SET s.title ="""+ '"' + str(df2['Title'][i]) + '"' + """
  RETURN s.stockcode, s.title
  """

  g.run(query)
```

Create a relationship between customers and stocks.

Since all the transactions are in the dataset, the relation is already known and present. To convert it into an RDS, cipher queries must be run in Neo4j to build the relationship.

```
# Storing transaction values in a list
transaction_list = df.values.tolist()
```

```
# storing all commands to build relationship in an empty list relation
relation = []
```

```
for i in transaction_list:
```

```
# the 9th column in df is customerID and 2nd column is stockcode which we
  are appending in the statement
  statement = """MATCH (a:customer),(b:stock) WHERE a.cid = """+'"' +
  str(i[8])+ '"' + """ AND b.stockcode = """ + '"' + str(i[1]) + '"' +
  """ CREATE (a)-[:bought]->(b) """
  relation.append(statement)

execute_transactions(relation)
```

Next, let's find similarities between users using the relationship created.

The Jaccard similarity can be calculated as the ratio between the intersection and the union of two sets. It is a measure of similarity, and as it is a percentage value, it ranges between 0% to 100%. More similar sets have a higher value.

```
def similar_users(id):
  # This query will find users who have bought stocks in common with the
    customer having id specified by user
  # Later we will find jaccard index for each of them
  # We wil return the neighbors sorted by jaccard index in descending order
    query = """
  MATCH (c1:customer)-[:bought]->(s:stock)<-[:bought]-(c2:customer)
  WHERE c1 <> c2 AND c1.cid ="""  + '"' + str(id) +'"' """
  WITH c1, c2, COUNT(DISTINCT s) as intersection
  MATCH (c:customer)-[:bought]->(s:stock)
  WHERE c in [c1, c2]
  WITH c1, c2, intersection, COUNT(DISTINCT s) as union
  WITH c1, c2, intersection, union, (intersection * 1.0 / union) as
  jaccard_index
  ORDER BY jaccard_index DESC, c2.cid
  WITH c1, COLLECT([c2.cid, jaccard_index, intersection, union])[0..15] as
  neighbors
  WHERE SIZE(neighbors) = 15    // return users with enough neighbors
  RETURN c1.cid as customer, neighbors
```

```
"""
    neighbors = pd.DataFrame([['CustomerID','JaccardIndex','Intersection',
    'Union']])
    for i in g.run(query).data():
    neighbors = neighbors.append(i["neighbors"])

    print("\n----------- customer's 15 nearest neighbors ---------\n")
    print(neighbors)
```

The following is a sample output.

```
similar_users('12347')
```

Figure 10-6 shows the output of users similar to customer 12347.

```
           ----------- customer's 15 nearest neighbors ---------

                   0            1             2       3
    0      CustomerID  JaccardIndex  Intersection   Union
    0           17396      0.111111            10      90
    1           13821      0.108333            13     120
    2           17097      0.107784            18     167
    3           13324      0.103093            10      97
    4           15658      0.099099            11     111
    5           15606      0.097345            11     113
    6           16389       0.09375             9      96
    7           18092      0.092784             9      97
    8           13814      0.091743            10     109
    9           13265       0.08871            11     124
    10          13488      0.087248            26     298
    11          12843      0.086957            12     138
    12          16618      0.086207            10     116
    13          15502      0.084821            19     224
    14          17722       0.08427            15     178
```

Figure 10-6. *The output*

```
similar_users(' 17975')
```

Figure 10-7 shows the output of users similar to customer 17975.

```
----------- customer's 15 nearest neighbors ---------

              0            1              2       3
    0   CustomerID  JaccardIndex   Intersection  Union
    0        15356      0.131098             43    328
    1        18231      0.126531             31    245
    2        14395      0.125436             36    287
    3        15856      0.124668             47    377
    4        16907      0.124424             27    217
    5        17787      0.123404             29    235
    6        15059       0.11236             30    267
    7        13344       0.11215             24    214
    8        16222      0.111111             29    261
    9        17085      0.108949             28    257
    10       17450       0.10687             28    262
    11       17865      0.106796             33    309
    12       16910      0.106061             35    330
    13       16549      0.105919             34    321
    14       13263      0.105155             51    485
```

Figure 10-7. *The output*

Now let's recommend the product based on similar users.

```
def recommend(id):
  # The query below is same as similar_users function
  # It will return the most similar customers
  query1 = """
  MATCH (c1:customer)-[:bought]->(s:stock)<-[:bought]-(c2:customer)
  WHERE c1 <> c2 AND c1.cid ="""" + '"' + str(id) +'"' """
  WITH c1, c2, COUNT(DISTINCT s) as intersection
  MATCH (c:customer)-[:bought]->(s:stock)
  WHERE c in [c1, c2]
  WITH c1, c2, intersection, COUNT(DISTINCT s) as union
  WITH c1, c2, intersection, union, (intersection * 1.0 / union) as
  jaccard_index
  ORDER BY jaccard_index DESC, c2.cid
  WITH c1, COLLECT([c2.cid, jaccard_index, intersection, union])[0..15]
  as neighbors
  WHERE SIZE(neighbors) = 15    // return users with enough neighbors
  RETURN c1.cid as customer, neighbors
```

```
"""
neighbors = pd.DataFrame([['CustomerID','JaccardIndex','Intersection',
'Union']])
neighbors_list = {}
for i in g.run(query1).data():
neighbors = neighbors.append(i["neighbors"])
neighbors_list[i["customer"]] = i["neighbors"]
print(neighbors_list)

# From the neighbors_list returned, we will fetch the customer ids of
those neighbors to recommend items
nearest_neighbors = [neighbors_list[id][i][0] for i in range
(len(neighbors_list[id]))]

# The below query will fetch all the items boughts by nearest neighbors
# We will remove the items which have been already bought by the target
customer
# Now from the filtered set of items, we will count how many times each
item is repeating within the shopping carts of nearest neighbors
# We will sort that list on count of repititions and return in
descending order
query2 = """
    // get top n recommendations for customer from their nearest
    neighbors
    MATCH (c1:customer),(neighbor:customer)-[:bought]->(s:stock)
    // all items bought by neighbors
    WHERE c1.cid = """ + '"' + str(id) + '"' """
      AND neighbor.cid in $nearest_neighbors
      AND not (c1)-[:bought]->(s)                    // filter for
      items that our user hasn't bought before

    WITH c1, s, COUNT(DISTINCT neighbor) as countnns // times bought by
    nearest neighbors
    ORDER BY c1.cid, countnns DESC
    RETURN c1.cid as customer, COLLECT([s.title, s.stockcode,
    countnns])[0..$n] as recommendations
    """
```

```
recommendations = pd.DataFrame([['Title','StockCode','Number of times
bought by neighbors']])
for i in g.run(query2, id = id, nearest_neighbors = nearest_neighbors,
n = 5).data():
#recommendations[i["customer"]] = i["recommendations"]
recommendations = recommendations.append(i["recommendations"])

# We will also print the items bought earlier by the target customer
print(" \n---------- Top 8 StockCodes bought by customer " + str(id) +
" -----------\n")

print(df[df['CustomerID']==id][['CustomerID','StockCode','Quantity']].
nlargest(8,'Quantity'))


bought = df[df['CustomerID']==id][['CustomerID','StockCode',
'Quantity']].nlargest(8,'Quantity')


print('\n-------Product Name of bought StockCodes ------\n')


print((df1[df1.StockCode.isin(bought.StockCode)]['Product Name']).
to_string())


# Here we will print the recommendations
print("------------ \n Recommendations for Customer {} -------
\n".format(id))
print(recommendations.to_string())
```

This function gets the following information.

- The top eight stock codes and product names bought by a particular customer

- Recommendations for the same customer and the number of times neighbors bought the same item

The following steps are followed to get to the desired.

1. Get the most similar customers for the given customer.

2. Fetch all the items bought by the nearest neighbors and remove the items the target customer has already bought.

3. From the filtered set of items, count the number of times each item is repeating within the shopping carts of nearest neighbors and then sort that list on the count of repetitions and return in descending order.

Now, let's try customer 17850.

```
recommend('17850')
```

Figure 10-8 shows the recommendations output for customer 17850.

```
{'17850': [['15497', 0.14814814814814814, 4, 27], ['17169', 0.14705882352941177, 5, 34], ['18170', 0.14285714285714285, 6, 4
2], ['15636', 0.1388888888888889, 5, 36], ['13187', 0.13725490196078433, 7, 51], ['15722', 0.1276595744680851, 6, 47], ['154
82', 0.11764705882352941, 4, 34], ['13161', 0.11538461538461539, 3, 26], ['14035', 0.11428571428571428, 4, 35], ['17866', 0.
1111111111111111, 3, 27], ['13884', 0.10714285714285714, 6, 56], ['14440', 0.10638297872340426, 5, 47], ['12747', 0.1, 5, 5
0], ['17742', 0.0967741935483871, 3, 31], ['14163', 0.09615384615384616, 5, 52]]}

---------- Top 8 StockCodes bought by customer 17850 -----------

      CustomerID StockCode  Quantity
285        17850    82494L        12
2629       17850    85123A        12
2634       17850     71053        12
2978       17850     71053        12
2983       17850    85123A        12
3299       17850     71053        12
3301       17850     21068        12
3302       17850    84029G        12

-------Product Name of bought StockCodes ------

135     Mediven Sheer and Soft 15-20 mmHg Thigh w/ Lac...
162     Heavy Duty Handlebar Motorcycle Mount Holder K...
179          AARCO Enclosed Wall Mounted Bulletin Board
669     3 1/2"W x 20"D x 20"H Funston Craftsman Smooth...
967     Awkward Styles Shamrock Flag St. Patrick's Day...
------------
 Recommendations for Customer 17850 -------
```

		0	1
2			
0		Title	StockCode Numbe
r of times bought by neighbors			
0	Puppy Apparel Clothes Clothing Dog Sweatshirt Pullover Coat Hoodie Gift for Pet		21754
5			
1	Mediven Sheer and Soft 15 20 mmHg Thigh w Lace Silicone Top Band CT Wheat II Ankle 8 8 75 inches		84879
5			
2	The Holiday Aisle LED C7 Faceted Christmas Light Bulb		22470
5			
3	Mediven Sheer and Soft 15 20 mmHg Thigh w Lace Silicone Top Band CT Wheat II Ankle 8 8 75 inches		82484
4			
4	MightySkins Skin Decal Wrap Compatible with Lifeproof Sticker Protective Cover 100 s of Color Options		22469
4			

Figure 10-8. *The output*

Next, let's try it on customer 12347.

```
recommend(' 12347')
```

Figure 10-9 shows the recommendations output for customer 12347.

{'12347': [['17396', 0.1111111111111111, 10, 90], ['13821', 0.10833333333333334, 13, 120], ['17097', 0.10778443113772455, 1
8, 167], ['13324', 0.1030927835015463, 10, 97], ['15658', 0.0990990990990991, 11, 111], ['15606', 0.09734513274336283, 11,
113], ['16389', 0.09375, 9, 96], ['18092', 0.09278350515463918, 9, 97], ['13814', 0.09174311926605505, 10, 109], ['13265',
0.08870967741935484, 11, 124], ['13488', 0.087248322147651, 26, 298], ['12843', 0.08695652173913043, 12, 138], ['16618', 0.0
8620689655172414, 10, 116], ['15502', 0.08482142857142858, 19, 224], ['17722', 0.08426966292134831, 15, 178]]}

```
---------- Top 8 StockCodes bought by customer 12347 -----------

       CustomerID StockCode  Quantity
99443       12347     23076       240
10526       12347     22492        36
99444       12347     22492        36
153949      12347     17021        36
200488      12347    84558A        36
10527       12347    85167B        30
10534       12347    84558A        24
43446       12347     84991        24

-------Product Name of bought StockCodes ------

33           Rosalind Wheeler Wall Mounted Bulletin Board
447      Eye Buy Express Kids Childrens Reading Glasses...
782      6pc Boy Formal Necktie Black & White Suit Set ...
1607     3 1/2"W x 32"D x 36"H Traditional Arts & Craft...
1820         Fruit of the Loom T-Shirts HD Cotton Tank Top
2668     Vickerman 14" Finial Drop Christmas Ornaments,...
-----------
 Recommendations for Customer 12347 -------


0           1                          2
0
Title  StockCode  Number of times bought by neighbors
0                             Handcrafted Ercolano Music Box Featuring Luncheon of the Boating Party by Ren
oir Pierre Auguste New YorkNew York    22697                 8
1                                                                                                    Window
Tint Film Mitsubishi all doors DIY     22698                 8
2   Girls Dress Up Kids Crafts Hair Kit With Hair Makes 10 Unique Hair Accessories Assortment of Kids Fashion Headbands Craf
t Kit Perfect Beauty Shop Play Date    22427                 6
3                             Elite Series Counter Height Storage Cabinet with Adjustable
Shelves 46 W x 24 D x 42 H Charcoal    23245                 6
4                                                                                         Port Authority K110 Dry Z
one UV Micro Mesh Polo Gusty Grey S    47566                 5
```

Figure 10-9. *The output*

Summary

This chapter briefly covered knowledge graphs and how graph-based recommendation engines work. You saw an actual implementation of an end-to-end graph-based recommender system using Neo4j knowledge graphs. The concepts used are very new and advanced and have become popular recently. Big players like Netflix and Amazon are shifting to graph-based systems for their recommendations; hence, the approach is very relevant and must-know.

## CHAPTER 11

Emerging Areas and Techniques in Recommender Systems

This book has shown you multiple implementations of recommender systems (also known as *recommendation systems*) using various techniques. You have gained a holistic view of all these methods. Topics like deep learning and graph-based approaches are still improving. Recommender systems have been a major research interest for a long time. Newer, more complex, and more interesting avenues have been discovered, and research continues in the same direction.

This chapter looks at real-time, context-aware, conversational, and multi-task recommenders to showcase the vast potential for research and growth in this field.

Real-Time Recommendations

In general, batch recommendations are computationally inexpensive and preferred because they can be generated daily (for example) and are much simpler to operationalize. But recently, more focus has been on developing real-time recommendations. Real-time recommendations are generally more computationally expensive since they must be generated on-demand and are based on live user interactions. Operationalizing real-time recommendations is also more complex.

Then why are real-time recommendations needed? They are essential when a time-based and mission-centric customer journey depends on context. In most scenarios, real-time demand needs to be met before the user loses interest and the demand fades. Also, real-time analysis of a customer's journey leads to better recommendations in the

© Akshay Kulkarni, Adarsha Shivananda, Anoosh Kulkarni, V Adithya Krishnan 2023
A. Kulkarni et al., *Applied Recommender Systems with Python*, https://doi.org/10.1007/978-1-4842-8954-9_11

current scenario. Conversely, a batch recommendation suggests products similar to those seen/bought by a customer in previous interactions.

Conversational Recommendations

Recent years have seen a lot of research and effort in developing more conversational systems. It is believed to revolutionize how human-computer interactions happen in the future. This influence can also be seen in the recent developments in recommender systems in the form of conversational systems.

Figure 11-1 shows the system design of conversational recommenders.

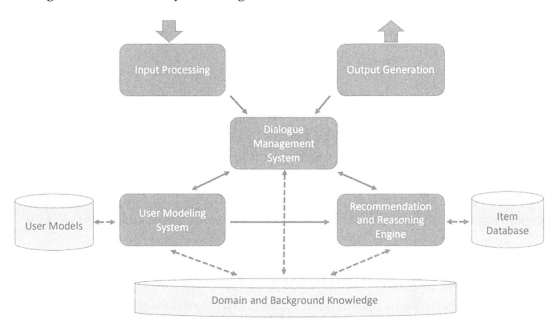

Figure 11-1. *System design of conversational recommendations*

Conversational recommender systems aim to produce recommendations from textual/spoken dialogs so that the users can interact with a computer in a natural, conversational manner. It has recently become extremely popular and is extensively used in voice assistants and chatbots. It uses natural language understanding (input) and generation (output). The different actions for various input dialogs are generated with the help of a dialog management system.

Context-Aware Recommenders

Researchers and practitioners have recognized the importance of contextual information in many fields, such as e-commerce personalization, information retrieval, ubiquitous and mobile computing, data mining, marketing, and management. Although there has been a lot of research in recommender systems, many existing approaches do not consider other contextual information, such as time, place, or other people's company, to find the most relevant. It focuses on recommending articles to users. (such as watching a movie or eating). There is a growing understanding that relevant contextual information is important in recommender systems and that it is important to consider it when making recommendations.

Context-aware recommender systems represent an emerging area of experimentation and research aimed at delivering more accurate content based on the user's context at any given moment. For example, is the user at home or on the go? Are they using a large or small screen? Morning or evening? Given the data available to a particular user, the contextual system may provide recommendations that the user will likely accept in these scenarios.

Figure 11-2 shows different types of contextual recommenders.

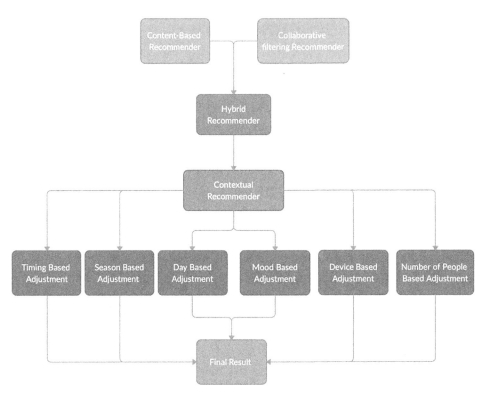

Figure 11-2. *Different types of contextual recommenders*

Multi-task Recommenders

In many domains, there are several rich and important sources of feedback to draw from while building a recommender system. For example, e-commerce websites generally record user visits (to product pages), user clicks (click-stream data), additions to carts, and purchases made at every user and item level. Post-purchase inputs like reviews and returns are also recorded.

Integrating these different forms of feedback is critical to building systems that yield better results rather than building a task-specific model. This is especially true where some data is sparse (purchases, returns, and reviews) and some data is abundant (such as clicks). In these scenarios, a joint model may use representations obtained from an abundant task to improve its predictions on a sparse one through a phenomenon known as *transfer learning*.

Multi-task learning is an ML learning approach in which multiple learning tasks are simultaneously tackled while exploiting their commonalities and differences. It has been predominantly used in NLP and computer vision, which has been quite successful. In recent times, the use of this method in building robust recommender systems has garnered a lot of interest. Building deep neural networks based on multi-task learning is widely used due to its multiple advantages.

- It avoids overfitting.

- It provides interpretable outputs to explain recommendations.

- It implicitly expands the data and thereby alleviates the sparsity problem.

Figure 11-3 explains the architecture of multi-task learning.

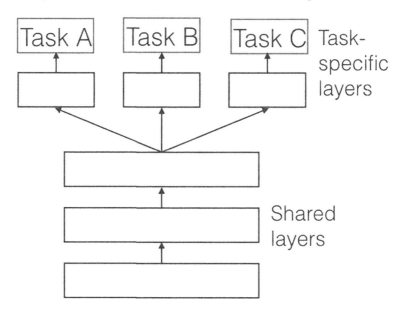

Figure 11-3. *Multi-task learning architecture*

You can also deploy multi-task learning to tackle cross-domain recommendations, where generating recommendations for each domain is a separate task.

Joint Representation Learning

Joint Representation Learning (JRL), a recent approach, is capable of learning multi-representation models of users and items simultaneously. It uses a deep representation learning architecture, where each type of information source (textual review, product images, rating points, etc.) is adopted to learn appropriate user and item representations.

Figure 11-4 explains the representation of JRL.

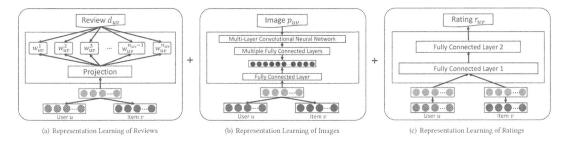

Figure 11-4. *Representation of JRL*

In JRL, multiple representations from different sources are integrated through a separate layer that obtains common representations for users and items. Finally, the per-source and joint representation layers are trained using pairwise learning to rank the top N recommendations. JRL is generally much quicker in online predictions than other deep learning approaches since it uses simple vector multiplications.

Conclusion

Recommender systems have been gaining traction since the start of the e-commerce era, but it has been around for some time. The first recommender system was developed in 1979 in a system called Grundy, a computer-based librarian that offered suggestions on books to read. The first commercial use of recommender systems was in the early 1990s. Since then, it has taken off because the financial incentives and time-saving qualities that recommender systems provide are unmatchable. Recommender systems have become essential for better user experience in many domains. The most popular example is that of Netflix and its recommendation engine, which receives heavy funding and focuses on research and development.

The constant need for recommender systems and their importance in various domains has led to a huge demand for building good, reliable, and robust systems. It calls for more research and innovation in developing these systems. This benefits businesses and helps users save time in making decisions and getting the most suitable option, which would be missed in most scenarios if done manually.

This book presented the various popular methods to implement an end-to-end recommender system in Python, ranging from basic arithmetic operations to advanced graph-based systems. All these methods can be useful depending on the requirements and the domain. Hands-on knowledge of these approaches will assist you in building the ideal Recommendation Engine (RC). We hope that this book is a useful tool for developers and practitioners. It should ease and further elevate the ongoing work and research in this exciting field by providing a deeper knowledge of the various concepts and implementation of recommendation engines.

Index

A

Accuracy, 195
Advanced machine learning concepts, 196
Apriori algorithm, 47–49
Area under the curve (AUC), 139, 140, 142
Artificial neural networks (ANN), 208
Association rule
    definition, 49
    mlxtend, implementation, 50, 51
    output, 50, 56
    visualization techniques, 54, 55, 57,
        58, 60–62

B

Backpropagation, 208
Best predictions, 125

C

Classification algorithm-based
    recommender system
    buying propensity model, 175
    data collection, 177
    DataFrame (pandas), 177, 178
    data preprocessing, 178–187
    decision tree, 198–200
    EDA, 188–192
    feature engineering, 187, 188
    implementation, 176
    KNN, 202–206

logistic regression, 194–198
random forest, 200–202
steps, 175
train-test split, 193
Classification algorithms, 206
Classification and regression
    tree (CART), 199
Clustering, 149
Clustering-based recommender systems
    approaches, 150
    data collection and
        downloading, 151
    data importing, 151, 152
    data preprocessing, 153
    elbow method, 159
    exploratory data analysis, 154–157
    hierarchical, 159
    implementation, 151
    k-means clustering, 158, 159
Co-clustering, 111, 120, 121, 123, 128
Collaborative-based filtering
    methods, 7, 129
Collaborative-based recommendation
    engines, 8
Collaborative filtering, 89
    approaches, 111, 112
    customer's purchase history and
        ratings, 89
    data1, 92, 93
    data collection, 91
    DataFrame, 91

243

© Akshay Kulkarni, Adarsha Shivananda, Anoosh Kulkarni, V Adithya Krishnan 2023
A. Kulkarni et al., *Applied Recommender Systems with Python*, https://doi.org/10.1007/978-1-4842-8954-9

Printed in the United States
by Baker & Taylor Publisher Services